SpringerBriefs in Applied Sciences and Technology

PoliMI SpringerBriefs

More information about this subseries at http://www.springer.com/series/11159
http://www.polimi.it

Maurizio Garrione · Filippo Gazzola

Nonlinear Equations for Beams and Degenerate Plates with Piers

POLITECNICO
MILANO 1863

Maurizio Garrione
Dipartimento di Matematica
Politecnico di Milano
Milan, Italy

Filippo Gazzola
Dipartimento di Matematica
Politecnico di Milano
Milan, Italy

ISSN 2191-530X   ISSN 2191-5318 (electronic)
SpringerBriefs in Applied Sciences and Technology
ISSN 2282-2577   ISSN 2282-2585 (electronic)
PoliMI SpringerBriefs
ISBN 978-3-030-30217-7   ISBN 978-3-030-30218-4 (eBook)
https://doi.org/10.1007/978-3-030-30218-4

This Springer imprint is published by the registered company Springer Nature Switzerland AG
The registered company address is: Gewerbestrasse 11, 6330 Cham, Switzerland

Preface

Many bridges have suffered unexpected oscillations both during construction and after inauguration, sometimes leading to collapse (see e.g. [2, 12]). Thanks to the videos available on the web [19], most people have seen the spectacular collapse of the Tacoma Narrows Bridge (TNB), which occurred in 1940: the torsional oscillations were considered the main cause of this dramatic event [3, 16]. Torsional oscillations leading to failures have also appeared in several other bridges. Examples include the collapses of the Brighton Chain Pier (1836), the Menai Strait Bridge (1839), the Wheeling Suspension Bridge (1854), and the Matukituki Suspension Footbridge (1977). We refer to [7, Chap. I] for more details and more historical events. Wide oscillations were also seen during the construction of the TNB and during the erection of the Storebaelt East Bridge (the second largest suspension bridge in the world) in 1998: on this occasion, it was necessary to fix some additional anchorages on the bottom of the sea [4, p. 32]. Let us emphasize that, although being generated by a different phenomenon (synchronization of pedestrians instead of vortex shedding), footbridges are also prone to display (lateral) oscillations; see again [7, Chap. I]. Here one may recall that such oscillations were seen on the very day of the opening of the London Millennium Bridge in 2000: the bridge had to be closed in order to prevent a possible tragedy [17]. This survey shows that new solutions are necessary to solve old problems and that unexpected oscillations still appear in bridges. These accidents raise some fundamental questions of deep mathematical interest, including how to improve the stability of suspension bridges. In order to analyze these phenomena from a theoretical point of view, we will use a wide variety of different skills.

Since we constantly refer to the main components of a suspension bridge throughout the book, let us explain in detail their roles (see Fig. 1). Four piers (or towers) sustain two parallel cables which, in turn, sustain the hangers. At their lower endpoints, the hangers are linked to the deck and sustain it from above. A suspension bridge is usually erected starting from the anchorages and the piers. Then the sustaining cables are installed between the two pairs of piers. Once the cables are in position, they furnish a stable working base from which the deck can be raised from floating barges. The hangers are hooked to the cables and the deck is

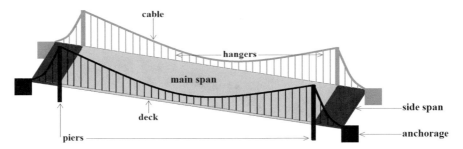

Fig. 1 Sketch of a suspension bridge

hooked to the hangers; this deforms the cable and stretches the hangers, which start their restoring action on the deck. We refer the reader to [15, Sect. 15.23] for further details. The part of the deck between the piers is called the main span while the parts outside the piers are called the side spans. The relative lengths of the three spans vary from bridge to bridge (see again [15]), and one of our purposes is precisely to establish whether there is an optimal position of the piers in order to maximize stability in a suitable sense. The complex composition of a suspension bridge yields two major difficulties: first, it is aerodynamically quite vulnerable; second, it appears very challenging to describe its behavior through simple and reliable mathematical models.

In particular, due to the large number of nonlinear interactions between its components, a fully reliable model appears out of reach. A typical example of this conflict can be found in the paper by Abdel-Ghaffar [1], who makes use of variational principles to obtain the combined equations of a suspension bridge motion in a fairly general nonlinear form. The effects of coupled vertical-torsional oscillations, as well as cross-distortional deformations of the stiffening structure, are described by separating them into four different kinds of displacements: the vertical displacement v, the torsional angle θ, the cross section distortional angle ψ, and the warping displacement u, which can be expressed in terms of θ and ψ. After making such a huge effort, Abdel-Ghaffar simplifies the problem by neglecting the cross section deformation, the shear deformation and the rotatory inertia; he obtains a coupled nonlinear vertical-torsional system of two equations in the two unknowns v and θ. These equations are finally linearized by neglecting terms which are considered small and the coupling effect disappears: see [1, 8, 9]. This procedure is quite common in the literature: first an attempt is made to describe the full model, then simplifications are proposed in order to deal with tractable equations. In this book, we follow the opposite procedure: first, we highlight the relevant phenomena in simple models, then we dress the models in order to make them as close as possible to a suspension bridge. We choose to model the deck of the bridge as a degenerate plate consisting of a beam with a continuum of cross sections free to rotate around the beam and we introduce a variety of possible nonlinearities. Due to the elastic behavior of several components in a bridge (in particular, the sustaining cables), one should also expect *nonlocal nonlinearities*. We study in detail the behavior of several nonlinearities in order to determine the ones that give responses in line with the phenomena visible in actual bridges.

A suspension bridge requires a fairly involved modeling and functional setting; the first instructive step towards a complete understanding of the different models describing its dynamics is the analysis of the linear stationary beam equation with piers. Such an equation fits within the framework of the so-called *multi-point* problems for ODEs, which were introduced in the pioneering work by Wilder [18] (see also [14] for some developments in the subsequent decades). As far as we are aware, evolution problems have almost never been studied in this setting (we mention the papers [10, 11] as the sole exceptions, partially related to the present investigation) and we insert within this framework also the degenerate plate, for which the multi-points become "multicross-sections". After exploring some physical motivations behind the considered models (Chap. 1), in Chap. 2, we briefly recall the basic framework for the linear stationary problem developed in [5], which will be useful for the study of the models presented in Chaps. 3 and 4. First, for the variational formulation we need to introduce suitable functional spaces. In Theorem 2.1, we recall that they are subspaces of codimension two of a second-order Sobolev space: the two missing dimensions prevent a nice regularity theory for weak solutions because the piers yield delta-type impulses on the beam. Also for the related eigenvalue problem, we cannot expect regularity of the eigenfunctions. In Theorems 2.3 and 2.4, we recall how to determine the eigenvalues of the linear problem, together with some of their properties on varying of the position of the piers; moreover, we recall the explicit form of the associated eigenfunctions, whose nodal properties are summarized in Theorem 2.5. As noted in the Federal Report [3], the placement and the number of zeros of the eigenfunctions are extremely important for the stability of bridges (see also the reproduction in Fig. 2.2, where a qualitative inventory of the oscillation modes seen at the TNB is drawn). The pattern of "zeros moving on the spans" is governed by several rules which can be described through eloquent pictures (see Figs. 2.1 and 2.3). Thanks to this functional framework we may dress the model and reach, in several steps, a model usable for the stability analysis of bridges.

The linear analysis, in particular, the behavior of the eigenvalues as the position of the piers varies, enables us to tackle the stability issue for some nonlinear evolution beam equations, for which we aim at determining the "best position" of the piers within the beam in order to maximize its stability. Motivated by both physical and mathematical arguments, we perform our analysis on *finite-dimensional* approximations of the phase space. We focus our attention on two different kinds of stability, the classical *linear stability*, the analysis of which is based on some Hill equations (see Definition 3.6), and a more "practical" stability that we introduced in [18] and that appears suitable for applications and nonlinear problems (see Definition 3.4). In the course of the analysis, we will show an existing connection between these two instabilities, the former being a clue to the latter. We restrict the instability analysis to a particular class of solutions of the nonlinear beam equation, namely those possessing a *prevailing mode*, i.e., for which the initial energy of the system is almost completely concentrated on a unique Fourier component (see Definition 3.2). These solutions well describe the behavior of actual bridges (see [3, p. 20] and Sect. 3.2). One is then interested in studying whether, for positive time, the energy remains mostly on the very same component or transfers towards other components. In other words, our stability analysis considers beams which initially oscillate close to a prevailing mode and explores whether this remains true in some interval of time. If not, we conclude that the prevailing mode is unstable and one is then interested in finding the *least energy* (or least amplitude of oscillation) which makes it unstable. The energy threshold for the stability of the beam is the least of these energies among all the possible prevailing modes. This means that, below this energy, the beam is stable and we expect the lower modes to have a greater likelihood of achieving this bound. Clearly, the energy threshold depends on the relative length of the side spans, and we are finally interested in finding the placement of the piers that maximizes it. As a by-product of our analysis, in Sect. 3.6, we identify the nonlinearities that make the beam behave more similarly to real bridges.

Finally, we complete the "model dressing" and focus our attention on the degenerate plate with a central beam and with cross sections free to rotate around it. By exploiting the results obtained for the simple beam, we make a precise choice of the nonlinearities to be inserted into the model. We analyze the impact of the two nonlinear restoring forces acting on the deck of a bridge, those exerted by the cables and the hangers (Sects. 4.3 and 4.4). A precise description of the stability is much more involved when the model is fully dressed, but we are still able to retrieve most of the phenomena visible for simple beams. The main targets are to study the impact of the hangers and of their elasticity on the stability thresholds, and to determine once again the optimal position of the piers that now do not constrain simple points but full segments (the cross sections between the piers). Some parts of our stability analysis are out of reach with purely theoretical tools and we use numerics. The scientific community is nowadays aware that many expensive experiments in wind tunnels may be replaced by numerics (see, e.g., the monograph [13], in particular its preface). And from [13, p. 13], we also take the motivation of their work: *we are not trying to substitute a designer with these optimization techniques, which would be impossible because of the complexity of real problems, but rather intending to help a designer not to fall into false steps that can be very probable for a design with great complexity*. A further goal of this book is precisely to introduce all the tools and nonlinearities needed for a systematic study of the stability in multi-point problems, in particular, those modeling suspension bridges. We conclude the book (Chap. 5) with a summary of the main results and their consequences, with some comments and some open problems, as well as possible future perspectives and developments.

Summarizing, in this book we provide the variational characterization of PDEs describing the stability of structures with internal piers, together with the proper functional setting for their study. As for the considered nonlinearities, we give strong hints about their impact on the stability analysis. We believe that all the necessary instruments to reach a complete understanding of the stability of suspension bridges are here introduced; our qualitative analysis should be seen as the starting point for a precise quantitative study of more complete models, taking into account the action of aerodynamic forces.

Milan, Italy   Maurizio Garrione
June 2019   Filippo Gazzola

Acknowledgements Both the authors are supported by the Gruppo Nazionale per l'Analisi Matematica, la Probabilità e le loro Applicazioni (GNAMPA) of the Istituto Nazionale di Alta Matematica (INdAM). The second author is also partially supported by the PRIN project Equazioni alle derivate parziali di tipo ellittico e parabolico: aspetti geometrici, disuguaglianze collegate, e applicazioni.

References

1. Abdel-Ghaffar AM (1982) Suspension bridge vibration: continuum formulation. J Eng Mech 108:1215–1232
2. Akesson B (2008) Understanding bridges collapses. CRC Press, Taylor & Francis Group, London
3. Ammann OH, von Kármán T, Woodruff GB (1941) The failure of the Tacoma Narrows Bridge. Federal Works Agency
4. de Miranda M, Petrequin M (1998) Storebaelt East Bridge: aspetti del montaggio e della realizzazione. Costruzioni Metalliche 6:27–42
5. Garrione M, Gazzola F (2017) Loss of energy concentration in nonlinear evolution beam equations. J Nonlinear Sci 27:1789–1827
6. Garrione M, Gazzola F (2020) Linear theory for beams with intermediate piers. Commun Contemp Math
7. Gazzola F (2015) Mathematical models for suspension bridges. Vol 15, MS&A, Springer
8. Gentile G, Mastropietro V, Procesi M (2005) Periodic solutions for completely resonant nonlinear wave equations with Dirichlet boundary conditions. Commun Math Phys 256:437–490
9. Gesztesy F, Weikard R (1996) Picard potentials and Hill's equation on a torus. Acta Math 176:73–107
10. Holubová G, Matas A (2003) Initial-boundary value problem for nonlinear string-beam system. J Math Anal Appl 288:784–802
11. Holubová G, Nečesal P (2010) The Fučík spectra for multi-point boundary-value problems. Electron J Differ Equ Conf 18:33–44
12. Imhof D (2004) Risk assessment of existing bridge structure. PhD Dissertation, University of Cambridge. See also http://www.bridgeforum.org/dir/collapse/type/ for the update of the Bridge failure database
13. Jurado JA, Hernández S, Nieto F, Mosquera A (2011) Bridge aeroelasticity, sensitivity analysis and optimal design. WIT Press, Southampton
14. Locker J (1973) Self-adjointness for multi-point differential operators. Pacific J Math 45:561–570
15. Podolny W (2011) Cable-suspended bridges. In: Brockenbrough RL, Merritt FS (eds) Structural steel designer's handbook: AISC, AASHTO, AISI, ASTM, AREMA, and ASCE-07 design standards, 5th edn. McGraw-Hill, New York

16. Scott R (2001) In the wake of Tacoma. Suspension bridges and the quest for aerodynamic stability. ASCE, Reston
17. Strogatz SH, Abrams DM, Eckhardt B, McRobie A, Ott E (2005) Theoretical mechanics: crowd synchrony on the Millennium Bridge. Nat Brief Communs 438:43–44
18. Wilder CE (1918) Problems in the theory of ordinary linear differential equations with auxiliary conditions at more than two points. Trans Am Math Soc 19:157–166
19. YouTube (1940) Tacoma Narrows Bridge collapse, video. http://www.youtube.com/watch?v=3mclp9QmCGs

Contents

About the Authors

Maurizio Garrione is Assistant Professor of Mathematical Analysis in the Department of Mathematics of the Politecnico di Milano, Italy. His research focus is in ordinary and partial differential equations, differential models and applications.

Filippo Gazzola is Professor of Mathematical Analysis in the Department of Mathematics of the Politecnico di Milano, Italy. His research focus is in partial differential equations in a broad sense, in calculus of variations and, in particular, in models for suspension bridges.

Chapter 1
The Physical Models

Abstract The physical models which will be considered throughout the book, describing the dynamics of beams and degenerate plates modeling suspension bridges, are introduced. They involve different kinds of nonlinear (nonlocal and local) energies of bending, stretching and displacement type behaving superquadratically. A physical interpretation of the different nonlinear terms considered is given.

Keywords Beams · Degenerate plates · Intermediate piers · Nonlocal energies · Restoring forces

We first describe the model for a doubly hinged beam divided in three adjacent spans (segments): the main (middle) span and two side spans separated by piers. Without loss of generality, we normalize the total length of the beam to 2π. Moreover, due to the extreme complexity of the asymmetric case, see [12], we restrict our attention to the case of symmetric beams, namely having equal side spans. We represent the beam as in Fig. 1.1.

The parameter $0 < a < 1$ determines the relative measure of the side spans with respect to the main span; the beam is hinged at the extremal points $\pm\pi$ and at the points $\pm a\pi$, corresponding to the position of the piers. The choice of considering symmetric beams is also motivated by the fact that most suspension bridges have equal side spans with

$$\frac{1}{2} \leq a \leq \frac{2}{3}, \tag{1.1}$$

see [20, Table 15.10].

If $\gamma > 0$ denotes the elastic restoring parameter, the linear beam equation reads

$$u_{tt} + u_{xxxx} + \gamma u = 0 \quad x \in I = (-\pi, \pi), \quad t > 0, \tag{1.2}$$

where u represents the vertical displacement; we will specify in Sect. 3.5 in which sense (1.2) is satisfied. We complement (1.2) with the boundary and internal conditions

$$u(-\pi, t) = u(\pi, t) = u(-a\pi, t) = u(a\pi, t) = 0 \quad t \geq 0, \tag{1.3}$$

© The Author(s), under exclusive license to Springer Nature Switzerland AG 2019
M. Garrione and F. Gazzola, *Nonlinear Equations for Beams and Degenerate Plates with Piers*, PoliMI SpringerBriefs,
https://doi.org/10.1007/978-3-030-30218-4_1

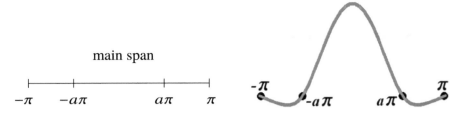

main span

Fig. 1.1 A beam with two symmetric piers and $0 < a < 1$

where the last two conditions state that no displacement occurs in correspondence of the piers $\{-a\pi, a\pi\}$ and give rise to a multi-point problem for (1.2). In the dynamics of the beam modeled by (1.2)–(1.3), the two internal conditions in the piers create an interaction between the three spans, otherwise the three segments would behave independently of each other.

Equation (1.2) may be solved by separation of variables but it is too far away from the final target of the present book, that is, modeling a bridge with a quantitative analysis of its instability with respect to the position of the piers. Two main ingredients are missing in (1.2): some nonlinearity which is intrinsic in complicated elastic structures such as bridges and the possibility of displaying torsional oscillations. We will reach a reliable bridge model in three steps: first we analyze the behavior of beams, then we study the response of different nonlinearities in order to detect the most reliable one, finally we turn to the analysis of a nonlinear degenerate plate.

As for the nonlinear term to be introduced in (1.2), a cubic nonlinearity naturally arises when large deflections of a beam or a plate are involved: in this case, the stretching effects suggest to use variants of the von Kármán theory [21], see also [10, 11] for a modern point of view and [13] for the adaptation of this theory to plates modeling bridges. In fact, when dealing with bridges, the nonlinearity should as well take into account the behavior of the sustaining cables and, for this reason, also the engineering literature deals with cubic nonlinearities, see e.g. [2, 3, 19].

Cubic nonlinearities within the equation arise when quartic energy terms are inserted into the model. We add to the bending energy the four different energies

$$\mathcal{E}_1(u) = \tfrac{1}{2}\int_I u_{xx}^2 + \tfrac{1}{4}\left(\int_I u_{xx}^2\right)^2, \ \mathcal{E}_2(u) = \tfrac{1}{2}\int_I u_{xx}^2 + \tfrac{1}{4}\left(\int_I u_x^2\right)^2,$$
$$\mathcal{E}_3(u) = \tfrac{1}{2}\int_I u_{xx}^2 + \tfrac{1}{4}\left(\int_I u^2\right)^2, \quad \mathcal{E}_4(u) = \tfrac{1}{2}\int_I u_{xx}^2 + \tfrac{1}{4}\int_I u^4,$$

of which the first three are nonlocal, while the last one is local. By inserting the kinetics, one finds that the total energy is $\mathcal{E}(u) = \tfrac{1}{2}\int_I u_t^2 + \mathcal{E}_i(u)$ $(i = 1, 2, 3, 4)$, so that the resulting evolution equations read, respectively:

$$u_{tt} + \left(1 + \|u_{xx}\|_{L^2}^2\right)u_{xxxx} = 0, \ u_{tt} + u_{xxxx} - \|u_x\|_{L^2}^2 u_{xx} = 0,$$
$$u_{tt} + u_{xxxx} + \|u\|_{L^2}^2 u = 0, \qquad u_{tt} + u_{xxxx} + u^3 = 0,$$

for $x \in I$ and $t > 0$. These equations are here written in strong form but, as we shall see, there is not enough regularity for this formulation: we instead stick to their weak form. This will be the object of Chap. 3. In all the energies $\mathcal{E}_i$ ($i = 1, 2, 3, 4$), the first (quadratic) term represents the bending energy while the quartic term has different physical meanings that we now discuss.

There are several reasons for considering $\mathcal{E}_1$. First, it may be derived from a variational principle of a stiffened beam with bending energy behaving superquadratically and nonlocally: this means that if the beam is bent somewhere, then this increases the resistance to further bending in all the other points. Moreover, as explained by Cazenave–Weissler [8, p. 110], we choose this energy in the hope that studying the behavior of the solution of the equation could provide an indication of what to expect from more complicated energies.

In 1950, Woinowsky-Krieger [22] modified the classical beam models by Bernoulli and Euler assuming a nonlinear dependence of the axial strain on the deformation gradient, by taking into account the stretching of the beam due to its elongation. Independently, Burgreen [6] obtained the very same nonlinear beam equation. This leads to the energy $\mathcal{E}_2$ from which the corresponding Euler–Lagrange equation can also be derived from a variational principle, now describing a stiffened beam with stretching energy behaving superquadratically and nonlocally: if the beam is stretched somewhere, then this increases the resistance to further stretching in all the other points (we refer to [14] for a damped and forced version). In spite of this analogy, we will see in Sect. 3.5 that the stretching energy *goes through the piers*, thereby behaving differently from the bending energy. More similar to $\mathcal{E}_1$ is the energy $\mathcal{E}_3$ which describes a stiffened beam where the displacement behaves superquadratically and nonlocally: if the beam is displaced from its equilibrium position in some point, then this increases the resistance to further displacements in all the other points. The corresponding equation is the fourth order version of the wave equation considered in [8, 9].

The energy $\mathcal{E}_4$ is more standard and represents a stiffened beam where the displacement behaves superquadratically but locally. The beam increases pointwise its resistance to displacements from the equilibrium position: in every point the resistance to displacement depends only on the displacement itself in the same point. The corresponding equation will be studied in Sect. 3.5. After a careful study of their behavior, we select the most reliable of these four nonlinearities: by "reliable" we mean here "well-describing the phenomena visible in real bridges".

The video [15] shows that there is *transmission of displacements* between spans. But if one seeks "tacoma narrows bridge collapse images" on Google, one finds pictures of the wide oscillations prior to the TNB collapse and one sees that, while a torsional motion is visible on the main span, the side spans do not display torsional displacements, see the qualitative reproduction in Fig. 1.2.

From a mathematical point of view, this means that

the matching between the displacements on the three spans

is in general not smooth.

Fig. 1.2 Qualitative behavior of the oscillations at the TNB the day of the collapse

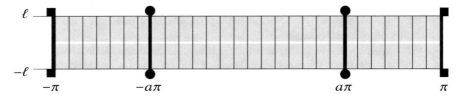

Fig. 1.3 A fish-bone model for a bridge with piers

This fact is confirmed by a careful look at the video [24] which clearly shows that, during the oscillations, the connection between the main span and the side spans *is not C^1*. In fact, Fig. 1.2, the pictures on the web, and the video also show that

within the deck only the displacement of the midline

is smooth during the oscillations.

This is why among several possible ways of modeling the deck of a bridge with piers, we choose to see it as a degenerate plate, composed by a beam representing the midline of the plate and by cross sections that are free to rotate around the beam, see Fig. 1.3.

This guarantees both a smooth midline displacement as in [15] and nonsmooth connections between spans. Not only this model appears particularly appropriate to describe the behavior of a real bridge, but it also enables us to exploit the analysis performed for the stability in beams. Again, we limit ourselves to consider the case of symmetric side spans; setting

$$I = (-\pi, \pi), \quad I_- = (-\pi, -a\pi), \quad I_0 = (-a\pi, a\pi), \quad I_+ = (a\pi, \pi),$$

the plate is identified with the planar rectangle

$$\Omega = I \times (-\ell, \ell) \subset \mathbb{R}^2, \quad (\ell > 0),$$

while the three spans are identified with

$$\Omega_0 := I_0 \times (-\ell, \ell) \quad \text{(main span)},$$

$$\Omega_- = I_- \times (-\ell, \ell), \quad \Omega_+ = I_+ \times (-\ell, \ell) \quad \text{(side spans)}.$$

The deck in actual bridges has a rectangular shape with two long edges of the order of 1 km and two shorter edges of the order of 14 m. Therefore, if we assume that the deck Ω has width 2ℓ and length 2π, after scaling this means that $\ell \approx \pi/75$. The white midline in Fig. 1.3 divides the roadway into two lanes and its vertical displacement is denoted by $u = u(x, t)$, for $x \in I$ and $t > 0$. The equilibrium position of the midline is $u = 0$, with

$$u > 0 \text{ corresponding to a } \textbf{downwards} \text{ displacement}. \qquad (1.4)$$

Each cross-section is free to rotate around the midline and its angle of rotation is denoted by $\alpha = \alpha(x, t)$. This model was called a *fish-bone* in [5].

The vertical displacements of the two endpoints of the cross sections (in position x and at time t) are given by

$$u(x, t) + \ell \sin \alpha(x, t) \quad \text{and} \quad u(x, t) - \ell \sin \alpha(x, t). \qquad (1.5)$$

Since we are only interested in describing accurately the behavior of the plate under small torsional angles α, the following approximations are legitimate:

$$\cos \alpha \cong 1 \quad \text{and} \quad \sin \alpha \cong \alpha. \qquad (1.6)$$

If we set $\theta = \ell \alpha$, this cancels the dependence on the width ℓ, see [5] for the details. In view of (1.6), the displacements (1.5) now read $u(x, t) \pm \theta(x, t)$. In a suspension bridge, these displacements generate forces due to the restoring action of the cables+hangers system at the endpoints of the cross sections; a crucial step is to find a reasonable form for these forces. We denote them by $f(u \pm \theta)$ and the corresponding potentials by $F(u \pm \theta)$, so that $F' = f$.

We derive the Euler–Lagrange equations for this structure using variational methods, as a consequence of an energy balance. Denoting by $M > 0$ the mass density of the beam, the kinetic energy of the beam is given by the well-known formula:

$$\frac{M}{2} \int_I u_t^2.$$

Since the plate is assumed to be homogeneous, the mass density of each cross section is also equal to $M > 0$. On the other hand, the kinetic energy of a rotating object is $\frac{1}{2} J \alpha_t^2$, where J is the moment of inertia and α_t is the angular velocity. The moment of inertia of a rod of length 2ℓ about the perpendicular axis through its center is given by $\frac{M\ell^2}{3}$. Hence, the kinetic energy of the rod having half-length ℓ, rotating around its center with angular velocity α_t, is given by

$$\frac{M}{6}\ell^2 \int_I \alpha_t^2 = \frac{M}{6} \int_I \theta_t^2.$$

Moreover, there exists a constant $\mu > 0$, depending on the shear modulus and on the moment of inertia of the pure torsion, such that the total potential energy of the cross sections is given by

$$\frac{\mu\ell^2}{2} \int_I \alpha_x^2 = \frac{\mu}{2} \int_I \theta_x^2.$$

The bending energy of the beam depends on its curvature: if $EI > 0$ is the flexural rigidity of the beam, it is given by

$$\frac{EI}{2} \int_I u_{xx}^2.$$

Finally, we derive the most delicate energy terms which create the *coupling* between the longitudinal displacement u and the torsional angle θ. Both the contributions of the cables and the hangers have to be taken into account.

• **Restoring force due to the cables**. As pointed out in the engineering literature (see, e.g., [3, 17]), the most relevant source of nonlinearity in suspension bridges, and hence the main contribution to instability, comes from the sustaining cables. The displacements of the endpoints $u \pm \theta$ of the cross sections of the deck stretch the cables which increase nonlinearly their tension and have a nonlocal restoring effect distributed throughout themselves and, in turn, on the deck. The tension at rest of the cables is proportional to the span/sag ratio. If a single beam is sustained by a cable having tension at rest $\gamma > 0$ and if the beam has a displacement w, then the term $\gamma \|w\|_{L^2}^2$ is a good measure for the geometric nonlinearity of the beam due to its displacement: as noticed in [7], the corresponding equation is a Hamiltonian system that combines certain properties of the linear beam equation and of the central force motion. The potential $G = G(w)$ reads

$$G(w) = \frac{\gamma}{4} \left(\int_I w^2 \right)^2. \tag{1.7}$$

We use here this principle to study the behavior of the edges $I \times \{\pm\ell\}$ of the cross sections of the plate $\Omega = I \times (-\ell, \ell)$. Therefore, the potential (1.7) is applied at both the endpoints of the cross sections. This means that w should be replaced by both $u \pm \theta$, giving rise to the potentials

$$G(u + \theta) = \frac{\gamma}{4} \left(\int_I (u + \theta)^2 \right)^2, \qquad G(u - \theta) = \frac{\gamma}{4} \left(\int_I (u - \theta)^2 \right)^2.$$

• **Restoring force due to the hangers**. The hangers have the delicate role to connect the cables and the deck, see Fig. 1 in preface. In some cases, they may be considered inextensible, see [17], but their main feature is that they can lose tension if the deck and the cables are too close to each other, a phenomenon called *slackening*, that was observed during the TNB collapse, see [1, V–12].

Several different forms for the restoring force due to the hangers have been suggested in literature, see [4] for a quick survey. At this stage, we simply denote the restoring force by f, postponing its explicit form to Sect. 4.4. The potentials $F = F(u \pm \theta)$ depend only on the *local* unknowns u, θ and not on their global behavior.

Thanks to all the above terms, we finally obtain the total energy associated with the fish-bone model:

$$E(u, \theta) = \frac{M}{2} \int_I u_t^2 + \frac{M}{6} \int_I \theta_t^2 + \frac{\mu}{2} \int_I \theta_x^2 + \frac{EI}{2} \int_I u_{xx}^2 + \frac{\gamma}{4} \left(\int_I (u + \theta)^2 \right)^2$$
$$+ \frac{\gamma}{4} \left(\int_I (u - \theta)^2 \right)^2 + \int_I F(u + \theta) + \int_I F(u - \theta).$$

This energy then leads to the following *nonlocal* system, for $x \in I$ and $t > 0$:

$$\begin{cases} Mu_{tt} + EIu_{xxxx} + 2\gamma \left(\int_I (u^2 + \theta^2) \right) u + 4\gamma \left(\int_I u\theta \right) \theta + f(u + \theta) + f(u - \theta) = 0 \\ \frac{M}{3} \theta_{tt} - \mu\theta_{xx} + 4\gamma \left(\int_I u\theta \right) u + 2\gamma \left(\int_I (u^2 + \theta^2) \right) \theta + f(u + \theta) - f(u - \theta) = 0. \end{cases} \quad (1.8)$$

If $\gamma = 0$ and f is linear, then system (1.8) decouples into two linear equations. The definition of *weak solution* of (1.8) will be given in Sect. 4.2.

In a slightly different setting, involving mixed space-time fourth order derivatives, a linear version of (1.8) with coupling terms was suggested by Pittel–Yakubovich [18], see also [23, p. 458, Chap. VI]. A nonlinear f was considered in (1.8) by Holubová–Matas [16], who were able to prove well-posedness for a forced-damped version of (1.8). Also in [5] the well-posedness of an initial-boundary value problem for (1.8) is proved for a wide class of nonlinear forces f. The fish-bone model described by (1.8), with *nonlinear f*, is able to display a possible transition between vertical and torsional oscillations within the main span: the former are described by u whereas the latter are described by θ.

To (1.8) we associate the boundary-internal-initial conditions

$$u(-\pi, t) = u(\pi, t) = \theta(-\pi, t) = \theta(\pi, t) = 0 \quad t \geq 0, \quad (1.9)$$

$$u(-a\pi, t) = u(a\pi, t) = \theta(-a\pi, t) = \theta(a\pi, t) = 0 \quad t \geq 0, \quad (1.10)$$

$$u(x, 0) = u_0(x), \; u_t(x, 0) = u_1(x), \; \theta(x, 0) = \theta_0(x), \; \theta_t(x, 0) = \theta_1(x) \quad x \in I. \quad (1.11)$$

The complete study of the stability for such a model, to be performed mainly through numerical tools, is the main scope of Chap. 4.

References

1. Ammann OH, von Kármán T, Woodruff GB (1941) The failure of the Tacoma Narrows Bridge. Federal Works Agency
2. Augusti G, Sepe V (2001) A "deformable section" model for the dynamics of suspension bridges. Part I: model and linear response. Wind Struct 4:1–18
3. Bartoli G, Spinelli P (1993) The stochastic differential calculus for the determination of structural response under wind. J Wind Eng Ind Aerodyn 48:175–188
4. Benci V, Fortunato D, Gazzola F (2017) Existence of torsional solitons in a beam model of suspension bridge. Arch Ration Mech Anal 226:559–585
5. Berchio E, Gazzola F (2015) A qualitative explanation of the origin of torsional instability in suspension bridges. Nonlinear Anal TMA 121:54–72
6. Burgreen D (1951) Free vibrations of a pin-ended column with constant distance between pin ends. J Appl Mech 18:135–139
7. Cazenave T, Haraux A, Weissler FB (1993) A class of nonlinear, completely integrable abstract wave equations. J Dyn Differ Equ 5:129–154
8. Cazenave T, Weissler FB (1995) Asymptotically periodic solutions for a class of nonlinear coupled oscillators. Port Math 52:109–123
9. Cazenave T, Weissler FB (1996) Unstable simple modes of the nonlinear string. Quart Appl Math 54:287–305
10. Chueshov I, Lasiecka I (2010) Von Kármán evolution equations. Well-posedness and long-time dynamics. Springer monographs in mathematics. Springer, New York
11. Ciarlet PG, Rabier P (1980) Les équations de von Kármán. Studies in mathematics and its applications 27. Springer, Berlin
12. Garrione M, Gazzola F (2020) Linear theory for beams with intermediate piers. Commun Contemp Math
13. Gazzola F, Wang Y (2015) Modeling suspension bridges through the von Kármán quasilinear plate equations, Contributions to nonlinear elliptic equations and systems, 269–297, Progr. Nonlinear Differential Equations Appl., 86, Birkhäuser/Springer
14. Ghisi M, Gobbino M, Haraux A (2018) An infinite dimensional Duffing-like evolution equation with linear dissipation and an asymptotically small source term. Nonlinear Anal Real World Appl 43:167–191
15. Hartman Mather B. https://a.msn.com/r/2/BBOvA55?m=nl-nl&referrerID=InAppShare&ocid=Nieuws (via Viral Hog, video)
16. Holubová G, Matas A (2003) Initial-boundary value problem for nonlinear string-beam system. J Math Anal Appl 288:784–802
17. Luco JL, Turmo J (2010) Effect of hanger flexibility on dynamic response of suspension bridges. J Eng Mech 136:1444–1459
18. Pittel BG, Yakubovich VA (1969) A mathematical analysis of the stability of suspension bridges based on the example of the Tacoma bridge (Russian). Vestnik Leningrad Univ 24:80–91
19. Plaut RH, Davis FM (2007) Sudden lateral asymmetry and torsional oscillations of section models of suspension bridges. J Sound Vib 307:894–905
20. Podolny W (2011) Cable-suspended bridges. In: Brockenbrough RL, Merritt FS (eds) Structural steel designer's handbook: AISC, AASHTO, AISI, ASTM, AREMA, and ASCE-07 design standards, 5th edn. McGraw-Hill, New York
21. von Kármán T (1910) Festigkeitsprobleme im maschinenbau. In: Encyclopaedia der Mathematischen Wissenschaften, vol IV/4 C. Springer, Leipzig, pp 348–352
22. Woinowsky-Krieger S (1950) The effect of an axial force on the vibration of hinged bars. J Appl Mech 17:35–36
23. Yakubovich VA, Starzhinskii VM (1975) Linear differential equations with periodic coefficients. Wiley, New York (Russian original in Izdat. Nauka, Moscow, 1972)
24. YouTube (1940) Tacoma Narrows Bridge collapse, video. http://www.youtube.com/watch?v=3mclp9QmCGs

Chapter 2
Functional Setting and Vibrating Modes for Symmetric Beams

Abstract The functional framework for beams with intermediate piers is introduced. First, some results for the associated linear stationary problem are given, showing the possible loss of regularity of the solutions, due to the presence of the piers. Then, a spectral result characterizing the eigenvalues is stated and the explicit form of the associated eigenfunctions is provided, together with their nodal properties.

Keywords Stationary problem · Multi-point conditions · Spectral analysis · Fundamental modes · Nodal characterization

In this chapter we consider a hinged beam of length 2π, represented by the segment $I = (-\pi, \pi)$, which has two intermediate piers in correspondence of the points $\pm a\pi$, $0 < a < 1$. We set

$$I_- = (-\pi, -a\pi), \qquad I_0 = (-a\pi, a\pi), \qquad I_+ = (a\pi, \pi),$$

so that $\overline{I} = \overline{I}_- \cup \overline{I}_0 \cup \overline{I}_+$. The aim is here to briefly recall the functional framework in which problem (1.2)–(1.3) can be settled and the vibrating modes of this kind of beam, determined in [2].

We start introducing the space

$$V(I) := \{u \in H^2 \cap H_0^1(I);\ u(\pm a\pi) = 0\}; \tag{2.1}$$

notice that the boundary and internal conditions

$$u(-\pi) = u(\pi) = u(-a\pi) = u(a\pi) = 0 \tag{2.2}$$

make sense since $V(I)$ embeds into $C^0(\overline{I})$. The space $V(I)$ can be characterized through the following statement [2, Theorem 1].

Theorem 2.1 *The space $V(I)$ is a subspace of $H^2 \cap H_0^1(I)$ having codimension 2, whose orthogonal complement is given by*

$$V(I)^\perp = \{v \in C^2(\overline{I}) \mid v(\pm\pi) = v''(\pm\pi) = 0,\ v'' \text{ is piecewise affine on } \overline{I}_-, \overline{I}_0 \text{ and } \overline{I}_+\}.$$

© The Author(s), under exclusive license to Springer Nature Switzerland AG 2019
M. Garrione and F. Gazzola, *Nonlinear Equations for Beams and Degenerate Plates with Piers*, PoliMI SpringerBriefs,
https://doi.org/10.1007/978-3-030-30218-4_2

As a consequence, we have that $V(I)^{\perp} \subset C^2(\overline{I})$, but $V(I)^{\perp} \cap C^3(I) = \{0\}$; in particular, the elements of $V(I)^{\perp}$ fail to be C^3 (except for the zero function) since each pier produces a discontinuity in the third derivative.

We now introduce the forced stationary version of (1.2). First, we recall that if there are no piers, the equation reads as

$$u''''(x) + \gamma u(x) = f(x) \qquad x \in I \tag{2.3}$$

and the natural functional space where solutions of (2.3) have to be sought is $H^2 \cap H_0^1(I)$, endowed with the scalar product $(u, v) \mapsto \int_I u'' v''$. The notion of weak solution is derived from a variational principle: the total energy $E(u)$ of the beam in position u is the sum of the bending energy, the restoring energy, and the forcing energy. If the beam does have intermediate piers, the energy is defined on the functional space $V(I)$, that is,

$$E(u) = \frac{1}{2} \int_I \left((u'')^2 + \gamma u^2 \right) - \langle f, u \rangle_V \qquad \forall u \in V(I), \tag{2.4}$$

where $\langle \cdot, \cdot \rangle_V$ denotes the duality pairing between $V(I)$ and $V'(I)$, its dual space. If $f \in L^1(I)$, then the duality product may be replaced by the integral $\int_I fu$. By computing the Fréchet derivative of E in $V(I)$, we obtain the following definition.

Definition 2.1 Let $f \in V'(I)$. We say that $u \in V(I)$ is a weak solution of (2.3)–(2.2) if

$$\int_I u'' v'' + \gamma \int_I uv = \langle f, v \rangle_V \qquad \forall v \in V(I). \tag{2.5}$$

In the next statement [2, Theorem 3 and Corollary 4] the regularity of weak solutions is discussed; we here denote by $\delta_{\pm a\pi} \in V'(I)$ the Dirac delta distributions at the points $-a\pi$ and $a\pi$.

Theorem 2.2 *Let $\gamma \geq 0$. For all $f \in V'(I)$ there exists a unique weak solution $u \in V(I)$ of (2.3)–(2.2), according to Definition 2.1. Moreover, if $f \in C^0(\overline{I})$, then:*
(i) the solution satisfies $u \in C^4(\overline{I}_-) \cap C^4(\overline{I}_0) \cap C^4(\overline{I}_+) \cap C^2(\overline{I})$ and $u''(\pm\pi) = 0$;
(ii) there exist $\alpha_f, \beta_f \in \mathbb{R}$ (depending on f, γ, a) such that $u'''' + \gamma u = f + \alpha_f \delta_{a\pi} + \beta_f \delta_{-a\pi}$ in distributional sense;
(iii) there exists a subspace $X(I) \subset C^0(\overline{I})$ of codimension 2 such that $u \in C^4(\overline{I})$ if and only if $f \in X(I)$;
(iv) we have that $u \in C^4(\overline{I})$ if and only if $u \in C^3(\overline{I})$; this occurs whenever u coincides with the unique classical solution U_f of the problem

$$U_f''''(x) + \gamma U_f(x) = f(x) \quad x \in I, \quad U_f(-\pi) = U_f(-a\pi) = U_f(a\pi) = U_f(\pi) = 0. \tag{2.6}$$

From Item (i) we see that a weak solution of (2.5) makes the beam globally hinged; this means that its least energy configuration is C^2 at the piers and displays no bending at the endpoints. This may lead to nonsmooth solutions of (2.5): indeed, Item (ii) says that

if the two-piers beam is subject to a continuous force f,

then each pier yields an additional load equal to some impulse depending on f.

Moreover, it was shown in [2, Sect. 2.2] that

the *upwards* displacement of the beam with piers is obtained with a

sign-changing load f pushing *downwards* close to the piers

and *upwards* far away from them.

Let us now recall the modes of vibration for the beam $I = (-\pi, \pi)$ with two intermediate piers. The eigenvalues μ and the corresponding eigenfunctions $e \in V(I)$ solve the eigenvalue problem in weak formulation

$$\int_I e''v'' = \mu \int_I ev \quad \forall v \in V(I). \tag{2.7}$$

One can show that any eigenfunction belongs to $C^2(\overline{I})$ and is of class C^∞ on each span I_-, I_0, I_+, and it necessarily satisfies the no-bending boundary conditions

$$e''(-\pi) = e''(\pi) = 0$$

at the endpoints of the beam, see [2, Sect. 3.1]. Since *all the eigenvalues are strictly positive*, we set $\mu = \lambda^4$ and seek $e \in V(I)$ such that

$$\int_I e''v'' = \lambda^4 \int_I ev \quad \forall v \in V(I). \tag{2.8}$$

The following statement holds, see [2, Theorem 6].

Theorem 2.3 *The set of all the eigenvalues $\mu = \lambda^4$ of (2.7) is completely determined by the values of $\lambda > 0$ such that*

$$\sin(\lambda\pi) \sinh(\lambda a\pi) \sinh[\lambda(1 - a)\pi] = \sinh(\lambda\pi) \sin(\lambda a\pi) \sin[\lambda(1 - a)\pi] \tag{2.9}$$

$$\cos(\lambda\pi) \cosh(\lambda a\pi) \sinh[\lambda(1 - a)\pi] = \cosh(\lambda\pi) \cos(\lambda a\pi) \sin[\lambda(1 - a)\pi]. \tag{2.10}$$

*In case (2.9), the corresponding eigenfunctions are **odd** and given by:*

- $\mathbf{O}_\lambda(x) = \sin(\lambda x)$ *if $\lambda \in \mathbb{N}$ (implying both $\lambda a \in \mathbb{N}$ and $\lambda(1 - a) \in \mathbb{N}$);*
- *the odd extension of*

$$\mathscr{O}_\lambda(x) = \begin{cases} \dfrac{\sinh[\lambda(1-a)\pi]}{\sinh(\lambda a\pi)} (\sinh(\lambda a\pi)\sin(\lambda x) - \sin(\lambda a\pi)\sinh(\lambda x)) & if\ x \in [0, a\pi] \\[2ex] \dfrac{\sin(\lambda a\pi)}{\sin[\lambda(1-a)\pi]} (\sin[\lambda(1-a)\pi]\sinh[\lambda(x-\pi)] - \sinh[\lambda(1-a)\pi]\sin[\lambda(x-\pi)]) & if\ x \in [a\pi, \pi] \end{cases}$$

if $\lambda \notin \mathbb{N}$ *(implying both $\lambda a \notin \mathbb{N}$ and $\lambda(1-a) \notin \mathbb{N}$).*

In case (2.10), *the corresponding eigenfunctions are* **even** *and given by*

- $\mathbf{E}_\lambda(x) = \cos(\lambda x)$ *if* $\lambda - 1/2 \in \mathbb{N}$ *(implying both $\lambda a - 1/2 \in \mathbb{N}$ and $\lambda(1-a) \in \mathbb{N}$);*
- *the even extension of*

$$\mathscr{E}_\lambda(x) = \begin{cases} \dfrac{\sinh[\lambda(1-a)\pi]}{\cosh(\lambda a\pi)} (\cosh(\lambda a\pi)\cos(\lambda x) - \cos(\lambda a\pi)\cosh(\lambda x)) & if\ x \in [0, a\pi] \\[2ex] \dfrac{\cos(\lambda a\pi)}{\sin[\lambda(1-a)\pi]} (\sinh[\lambda(1-a)\pi]\sin[\lambda(\pi-x)] - \sin[\lambda(1-a)\pi]\sinh[\lambda(\pi-x)]) & if\ x \in [a\pi, \pi] \end{cases}$$

if $\lambda - 1/2 \notin \mathbb{N}$ *(implying both $\lambda a - 1/2 \notin \mathbb{N}$ and $\lambda(1-a) \notin \mathbb{N}$).*

The eigenfunctions $\mathbf{O}_\lambda$ and $\mathbf{E}_\lambda$ are of class C^∞ and satisfy the strong form of the eigenvalue problem

$$e'''' = \lambda^4 e.$$

On the other hand, $\mathscr{O}_\lambda$ and $\mathscr{E}_\lambda$ are only C^2; by computing explicitly their third derivative one can formally write Eq. (2.8) in strong form, as in Item (ii) of Theorem 2.2. Precisely, we have

$$e'''' = \lambda^4 e + \alpha_\lambda \delta_{a\pi} + \beta_\lambda \delta_{-a\pi},$$

for suitable constants α_λ and β_λ. It is also of crucial importance to notice that the eigenfunctions provided by Theorem 2.3 are orthogonal in $L^2(I)$ and in $V(I)$, but they *are not orthogonal* in $H^1(I)$, a fact that will have profound consequences when dealing with nonlinear models containing terms coming from the stretching energy $\mathscr{E}_2$ mentioned in Chap. 1, see Chap. 3.

The eigenvalue curves implicitly defined by (2.9) and (2.10) in the plane (a, λ) are depicted in Fig. 2.1; they fulfill the following properties, according to [2, Theorem 8].

Theorem 2.4 *For any $a \in (0, 1)$, the eigenvalues $\mu = \lambda^4$ of problem (2.8) are simple and form a countable set, the corresponding eigenfunctions are of class C^2 and form an orthogonal basis of $V(I)$. Moreover, (2.9) and (2.10) implicitly define, for $a \in (0, 1)$, a family of analytic functions $a \mapsto \lambda(a)$.*

It turns out that, even if all the eigenvalues are simple, the spectral gaps can be very small. This means that the corresponding modes of the linear evolution equation (1.2) have fairly similar frequencies. Notice moreover that the curves given by (2.9) are symmetric with respect to $a = 1/2$, since (2.9) is invariant upon the substitution $[a \to 1-a]$. In the case $a = 1/2$, Theorem 2.3 takes indeed the following simpler form [2, Corollary 7].

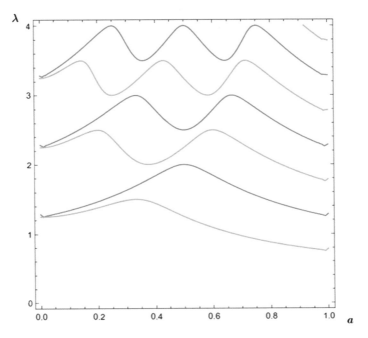

Fig. 2.1 The curves implicitly defined by (2.9)–(2.10) in the region $(a, \lambda) \in (0, 1) \times (0, 4)$

Corollary 2.1 *Let $a = 1/2$. The eigenvalues $\mu = \lambda^4$ of (2.7) are completely determined by the values of $\lambda > 0$ such that*

$$\sin(\lambda \pi/2) = 0 \quad or \quad \tan(\lambda \pi/2) = \tanh(\lambda \pi/2) \quad or \quad \tan(\lambda \pi) = \tanh(\lambda \pi).$$

In the first case, a corresponding eigenfunction is given by $\mathbf{O}_\lambda(x) = \sin(\lambda x)$, while in the other two cases it is given, respectively, by the odd extension of

$$\mathcal{O}_\lambda(x) = \begin{cases} \dfrac{\sin(\lambda x)}{\sin(\lambda \pi/2)} - \dfrac{\sinh(\lambda x)}{\sinh(\lambda \pi/2)} & \text{if } x \in [0, \pi/2] \\[2mm] \dfrac{\sinh[\lambda(x - \pi)]}{\sinh(\lambda \pi/2)} - \dfrac{\sin[\lambda(x - \pi)]}{\sin(\lambda \pi/2)} & \text{if } x \in [\pi/2, \pi], \end{cases}$$

and by the even extension of

$$\mathcal{E}_\lambda(x) = \begin{cases} \tanh\left(\dfrac{\lambda \pi}{2}\right)\left(\cosh\left(\dfrac{\lambda \pi}{2}\right)\cos(\lambda x) - \cos\left(\dfrac{\lambda \pi}{2}\right)\cosh(\lambda x)\right) & \text{if } x \in [0, \pi/2] \\[2mm] \cotan\left(\dfrac{\lambda \pi}{2}\right)\left(\sinh\left(\dfrac{\lambda \pi}{2}\right)\sin[\lambda(\pi - x)] - \sin\left(\dfrac{\lambda \pi}{2}\right)\sinh[\lambda(\pi - x)]\right) & \text{if } x \in [\pi/2, \pi]. \end{cases}$$

In order to properly study the instability phenomena for the oscillations of a suspension bridge, the placement of the zeros of the vibrating modes is of crucial importance. This was already noticed in the Federal Report [1], see the reproduction in Fig. 2.2 where an inventory of the modes of oscillation seen at the TNB is

Fig. 2.2 Zeros seen at the
TNB: hand reproduction of
Drawing 4 in [1]

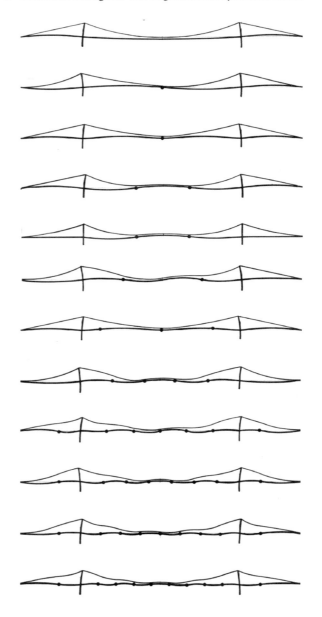

drawn. We here recall some results in this direction; to state them, notice that, for a
given $a \in (0, 1)$, Theorem 2.4 allows us to sort the eigenvalues in increasing order
$\{\lambda_0, \lambda_1, \lambda_2, \ldots\}$ and to label the associated eigenfunctions as $\{e_0, e_1, e_2, \ldots\}$. We will
speak about *even* and *odd* eigenfunctions and eigenvalues, referring to such labels.
In [2, Theorem 10], the eigenfunctions e_n were classified according to their nodal

properties. To properly define the number of zeros of an eigenfunction, it has to be noticed that an eigenfunction always vanishes in correspondence of the piers and such zeros may also be double; Corollary 2.1 provides an example ($a = 1/2$) where the eigenfunctions $\mathcal{O}_\lambda$ of (2.8) have this feature. If the eigenfunction has a double zero at the piers, its restriction to the central span is clamped, while its restrictions to the side spans are partially hinged and partially clamped. However, the zeros occurring in correspondence of the piers cannot have higher multiplicity; moreover, double zeros may occur only in correspondence of the piers. It is thus possible to define the index of an eigenfunction e_λ in I by

$$i(e_\lambda) = \begin{cases} \#\{x \in I_- \cup I_0 \cup I_+ \mid e_\lambda(x) = 0\} & \text{if } e'_\lambda(a\pi) \neq 0 \\ \#\{x \in I_- \cup I_0 \cup I_+ \mid e_\lambda(x) = 0\} + 2 & \text{if } e'_\lambda(a\pi) = 0. \end{cases}$$

Hence, double zeros at the piers are counted as two additional simple zeros.

We can now recall the complete description of the placement of the zeros of the eigenfunctions given in [2, Theorem 10]; fixed $a \in (0, 1)$, we here denote the n-th corresponding eigenvalue by $\lambda_n(a)$ and the associated eigenfunction by $e_{\lambda_n(a)}$, $n = 0, 1, 2, \ldots$, in order to underline the dependence of the eigenvalues and of the eigenfunctions on a. Moreover, we introduce the sequence $\{\Lambda_n\}_n$ whose elements are implicitly defined as follows:

$$\tan(\Lambda_{2k}\pi) = -\tanh(\Lambda_{2k}\pi), \qquad \tan(\Lambda_{2k+1}\pi) = \tanh(\Lambda_{2k+1}\pi).$$

It can be seen that $\{\Lambda_n\}_n$ coincides with the spectrum of the clamped eigenvalue problem (on $H_0^2(I)$).

Theorem 2.5 *For $a \in (0, 1)$, it holds that $i(e_{\lambda_n(a)}) = n$, for every $n = 0, 1, 2, \ldots$. Fixed an integer $n \geq 0$, on decreasing of a the zeros of $e_{\lambda_n(a)}$ move by couples from the central span to the side spans whenever the curve $\lambda = \lambda_n(a)$ intersects one of the hyperbolas $\{\lambda = \Lambda_k/a\}$, for some integer $k \geq 0$ having the same parity as n.*

By looking at Fig. 2.3, we see that the hyperbolas $\{\lambda = \Lambda_k/a\}$ describe a countable set of lines, each of which intersects the countable set of curves representing the eigenvalues in a countable number of points. Therefore, double zeros in the piers are possible only for a countable set of values of $a < 1$, that is,

for almost every $0 < a < 1$ all the eigenfunctions have simple zeros in the piers.

Theorem 2.5 states that a double zero is placed in a pier (and by symmetry also in the other one) each time that the curve $(a, \lambda_n(a))$ crosses the graph of one of the hyperbolas $\{\lambda = \Lambda_k/a\}$ (with k having the same parity as n). From there on, proceeding in the direction of decreasing a, the zeros of $e_{\lambda_n(a)}$ move from I_0 to I_+ (and I_-); for any odd n and for a sufficiently small so that $\lambda_n(a)$ lies below the hyperbola $\{\lambda = \Lambda_1/a\}$, all the zeros of $e_{\lambda_n(a)}$ lie in the lateral spans, except for the zero in the origin. For even n, the threshold becomes $\lambda_n(a) < \Lambda_0/a$ and no zeros of

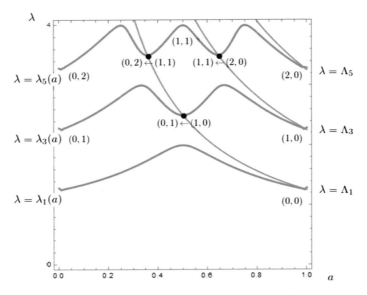

Fig. 2.3 A visual description of Theorem 2.5 for the curves $\lambda = \lambda_{2m+1}(a)$

$e_{\lambda_n(a)}$ at all belong to I_0 below this threshold. This pattern is represented for some odd eigenfunctions in Fig. 2.3, where the numbers (α, β) in parentheses denote, respectively, the number of zeros of the eigenfunction in $(0, a\pi)$ and in $(a\pi, \pi)$. The sum $\alpha + \beta$ is constant on each branch.

We also recall a result regarding the asymptotic behavior of the eigenvalues, which will become useful in Sect. 3.4.3 (see [2, Theorem 11]).

Theorem 2.6 *For every $a \in (0, 1)$, any interval of width 3 contains at least three values of λ for which $\mu = \lambda^4$ is an eigenvalue of (2.8). As a consequence,*

$$\lim_{n \to +\infty} \frac{\lambda_{n+1}(a)}{\lambda_n(a)} = 1 \qquad \forall a \in (0, 1). \tag{2.11}$$

In particular, it can be seen that

$$\lambda_{2m+1} \in [m + 1, m + 2], \tag{2.12}$$

a fact which will be useful to prove Theorem 3.3 below.

Finally, in Figs. 2.4 and 2.5 we plot the first twelve eigenfunctions for $a = 14/25$ (the ratio of the spans of the TNB, according to [1]) and $a = 1/2$, indicating with $\bullet$ the position of the piers.

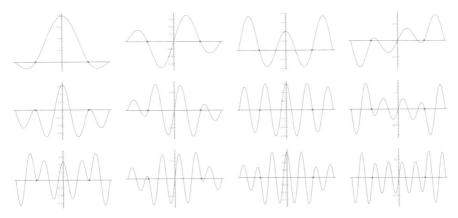

Fig. 2.4 The first twelve L^2-normalized eigenfunctions of (2.7) when $a = 14/25$

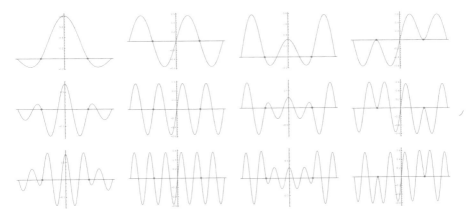

Fig. 2.5 The first twelve L^2-normalized eigenfunctions of (2.7) when $a = 1/2$

References

1. Ammann OH, von Kármán T, Woodruff GB (1941) The failure of the Tacoma Narrows Bridge. Federal Works Agency
2. Garrione M, Gazzola F (2020) Linear theory for beams with intermediate piers. Commun Contemp Math

Chapter 3
Nonlinear Evolution Equations for Symmetric Beams

Abstract The stability of solutions with a prevailing mode in different nonlinear models for beams with intermediate piers is studied. Both linear stability (for bimodal solutions) and a suitable notion of nonlinear stability are investigated, introducing a proper concept of energy threshold of instability and then determining the optimal placement of the piers, leading to the highest energy threshold. Moreover, it is shown that the nonlinearity which better describes the behavior of actual bridges, among those considered, is the one where the displacement behaves superquadratically and nonlocally.

Keywords Beams · Evolution problem · Prevailing modes · Stability · Optimal placement of the piers

We consider here a symmetric beam subject to different nonlinear forces. According to suitable definitions which will be given in the course, we study the stability of solutions with a prevailing mode. The final purposes are to identify the nonlinearity better describing the behavior of real structures and to locate the position of the piers maximizing the stability.

3.1 From Linear to Nonlinear Beam Equations

The general nonlinear equations describing our models read

$$u_{tt} + u_{xxxx} + \gamma_1 \|u_{xx}\|_{L^2}^2 u_{xxxx} - \gamma_2 \|u_x\|_{L^2}^2 u_{xx} + \gamma_3 \|u\|_{L^2}^2 u + f(u) = 0 \quad (3.1)$$

and take into account nonlocal factors of bending and stretching type and local restoring forces. Equation (3.1) is considered together with the initial conditions

$$u(x, 0) = u_0(x), \quad u_t(x, 0) = u_1(x) \quad x \in I. \tag{3.2}$$

We focus our attention on the "coercive" case where

© The Author(s), under exclusive license to Springer Nature Switzerland AG 2019
M. Garrione and F. Gazzola, *Nonlinear Equations for Beams and Degenerate Plates with Piers*, PoliMI SpringerBriefs,
https://doi.org/10.1007/978-3-030-30218-4_3

$$\gamma_1, \gamma_2, \gamma_3 \geqslant 0, \qquad f \in \mathrm{Lip}_{\mathrm{loc}}(\mathbb{R}), \qquad f(s)s \geqslant 0 \quad \forall s \in \mathbb{R}, \tag{3.3}$$

although the setting discussed in the forthcoming sections may be extended to more general nonlinearities. Note that (3.3) includes the case where $\gamma_1 = \gamma_2 = \gamma_3 = 0$ and $f(s) = \gamma s$, namely the linear equation (1.2), whose stationary solutions have been discussed in Chap. 2. One of our purposes is to examine the impact of each nonlinearity on the stability analysis: this will be done in Sect. 3.6.

Let us make precise what is meant by weak solution of (3.1).

Definition 3.1 We say that $u \in C^0(\mathbb{R}_+; V(I)) \cap C^1(\mathbb{R}_+; L^2(I)) \cap C^2(\mathbb{R}_+; V'(I))$ is a weak solution of (3.1)–(3.2) if $u_0 \in V(I)$, $u_1 \in L^2(I)$ and, for all $v \in V(I)$ and $t > 0$, one has

$$\langle u_{tt}, v \rangle_V + (1 + \gamma_1 \|u_{xx}\|_{L^2}^2) \int_I u_{xx} v'' + \gamma_2 \|u_x\|_{L^2}^2 \int_I u_x v' + \gamma_3 \|u\|_{L^2}^2 \int_I uv + \int_I f(u)v = 0.$$

We recall that $u(t) \in V(I)$ implies the conditions

$$u(-\pi, t) = u(\pi, t) = u(-a\pi, t) = u(a\pi, t) = 0 \qquad t \geqslant 0. \tag{3.4}$$

Existence and uniqueness of a weak solution are somehow straightforward.

Proposition 3.1 *Assume* (3.3). *There exists a unique weak solution u of* (3.1)–(3.2) *provided that:*
either $\gamma_1 = 0$, $u_0 \in V(I)$ and $u_1 \in L^2(I)$;
or $\gamma_1 > 0$ and $u_0 \in V(I)$, $u_1 \in L^2(I)$ have a finite number of nontrivial Fourier coefficients.
Moreover, $u \in C^2(\overline{I} \times \mathbb{R}_+)$ and $u_{xx}(-\pi, t) = u_{xx}(\pi, t) = 0$ for all $t > 0$.

By Theorem 2.4, the proof can be carried on with a Galerkin method, since the assumptions (3.3) yield a coercive stationary problem: it can be obtained by combining arguments from [16, Theorem 1] (local case) and [13] (nonlocal case). The case $\gamma_1 > 0$ needs stronger assumptions on the initial data because the Galerkin procedure does not allow to control the related nonlocal term. Due to the widely discussed lack of regularity at the piers, see Chap. 2, one cannot expect the solutions to be strong or classical. One reaches the C^2-regularity by arguing as in [23, Lemma 2.2], see also the proof of [17, Theorem 3].

In order to simplify our analysis, we consider the eigenfunctions of problem (2.7) normalized in L^2, sorting them according to the order of the corresponding (simple) eigenvalues, as in Theorem 2.5:

$$\{e_n\}_{n \in \mathbb{N}}, \qquad \|e_n\|_{L^2} = 1.$$

The initial data associated with (3.1) may be expanded in Fourier series with respect to the basis $\{e_n\}_n$:

$$u(x, 0) = \sum_{n \in \mathbb{N}} \alpha_n e_n, \quad u_t(x, 0) = \sum_{n \in \mathbb{N}} \beta_n e_n. \tag{3.5}$$

The study of (3.1) requires to modify some features of linear problems that are lost when dealing with nonlinear problems; let us briefly comment about this point by restricting our attention to a particular example. If in (3.1) we take $\gamma_1 = \gamma_2 = \gamma_3 = 0$ and $f(u) = \varepsilon u^3$ for some $\varepsilon \geqslant 0$, then Definition 3.1 states that a weak solution satisfies

$$\langle u_{tt}, v \rangle_V + \int_I u_{xx} v'' + \varepsilon \int_I u^3 v = 0 \qquad \forall v \in V(I), \quad t > 0. \tag{3.6}$$

We take initial conditions concentrated on a sole mode, namely

$$u(x, 0) = \alpha e_n(x), \quad u_t(x, 0) = 0 \qquad x \in I, \tag{3.7}$$

for some $\alpha > 0$ and some e_n eigenfunction of (2.8) with associated eigenvalue $\mu_n = \lambda_n^4$. In the *linear case* $\varepsilon = 0$, the solution of (3.6)–(3.7) is given by $u(x, t) = \alpha \cos(\lambda_n^2 t) e_n(x)$, which is periodic-in-time and proportional to the initially excited mode e_n.

But the nonlinear case $\varepsilon > 0$ is extremely more complicated and a full understanding of it is still missing. The milestone paper by Rabinowitz [30] started the systematic study of the existence of periodic solutions for the nonlinear wave equation

$$u_{tt} - u_{xx} + f(x, u) = 0 \quad x \in (0, \pi), \quad t > 0, \qquad u(0, t) = u(\pi, t) = 0 \quad t \geqslant 0. \tag{3.8}$$

Prior to [30], only perturbation techniques were available and the nonlinearity was assumed to be small in some sense. Rabinowitz proved that, under suitable assumptions on the nonlinearity f, this equation admits periodic solutions of any rational period. More recent works, see e.g. [7, 19], imply that there exist T-periodic solutions of (3.8) for all T belonging to sets of almost full measure. Periodic solutions were also found for the related nonlinear problem

$$u_{tt} + u_{xxxx} + f(x, u) = 0 \quad x \in (0, \pi), \quad t > 0,$$

$$u(0, t) = u(\pi, t) = u_{xx}(0, t) = u_{xx}(\pi, t) = 0 \quad t \geqslant 0,$$

for a beam without piers, see [26–28]. For this equation, periodic solutions of given period appear difficult to be determined, see [3] for some computer assisted results on this topic. Indeed, it is no longer true that a solution with initial conditions (3.7) is proportional to the mode e_n, because the energy immediately spreads on infinitely many other modes. A further difference between linear and nonlinear equations is that the latter may display some *instability*, meaning by this that small perturbations may lead to huge differences in the outcomes (in a suitable sense).

In order to overcome these difficulties, we argue as follows. First, we remark that the two parameters ε and α—respectively measuring the strength of the nonlinearity in (3.6) and the initial amplitude in (3.7)—play a similar role. To see this, set $w = \sqrt{\varepsilon} u$

in (3.6), so that w satisfies

$$
\begin{cases}
\langle w_{tt}, v \rangle_V + \int_I w_{xx} v'' + \int_I w^3 v = 0 & \forall v \in V(I), \quad t > 0 \\
w(x, 0) = \alpha \sqrt{\varepsilon}\, e_n(x), \quad w_t(x, 0) = 0 & x \in I,
\end{cases}
\tag{3.9}
$$

where the equation is independent of ε and the initial amplitude is now $\alpha \sqrt{\varepsilon}$. Therefore, if either α or ε is small, we are close to a linear regime and (3.6), as well as (3.9), can be tackled with perturbative methods and KAM theory (see, e.g., [24, 35]). But (3.9) also shows that for increasing α and ε the equation will behave more and more nonlinearly.

The second step is based on the simple remark that if one chooses initial data yielding a periodic solution of (3.6), then the solution does not spread energy since it coincides with the periodic solution. But, as we already mentioned, it is not easy to find explicit periodic solutions. Therefore, we prefer to still focus on the initial conditions (3.7), which give rise to a periodic solution for $\varepsilon = 0$, and study in detail how the energy of (3.6) initially concentrated on the mode e_n spreads among the other modes. In Sect. 3.3, we show that this may happen in two different ways, which need to be characterized sharply in order to define the stability of the beam. It turns out that two main ingredients influence the instability: the initially excited mode e_n and its amplitude α. Our purpose is to study the instability depending on the mode, on its initial amplitude, and on the position of the piers. Then we aim at finding the placement of the piers maximizing the stability in a suitable sense: this will be done in Sects. 3.4 and 3.5.

3.2 Prevailing Modes and Invariant Subspaces

The notion of instability we are interested in, is restricted to a particular class of solutions of (3.1). This class is motivated by the behavior of actual bridges. From the Report on the TNB collapse [1, p.20] we learn that, in the months prior to the collapse,

**one principal mode of oscillation prevailed
and the modes of oscillation frequently changed.**

This means that, even if the oscillations were governed by (disordered) forcing and damping, the oscillation itself was quite simple to describe, see also the sketch in Fig. 2.2. We characterize the solutions of (3.1) possessing a *prevailing mode* as the solutions having initial data concentrating most of the initial energy on a sole mode.

Definition 3.2 Let $0 < \eta < 1$. We say that a weak solution of (3.1)–(3.2), with initial data (3.5), has the j-th mode η-*prevailing* if j is the only integer for which:

$$
\sum_{n \neq j} (\alpha_n^2 + \beta_n^2) < \eta^4 (\alpha_j^2 + \beta_j^2).
\tag{3.10}
$$

For this solution, all the other modes $k \neq j$ are called *residual*. We denote by $\mathbb{P}_j$ the set of all the solutions of (3.1) with η-prevailing mode j.

We will comment on the role of the parameter η in Sect. 3.3. By now, we only anticipate that a reasonable choice of η is given by $\eta = 0.1$; from a physical point of view, this highlights an initial difference between prevailing and residual modes of two orders of magnitude. In Table 3.13 in Sect. 3.7 we also consider other values of η. Obviously, not all the solutions of (3.1) have an η-prevailing mode. In the sequel, we restrict our attention to initial conditions (3.5) without kinetic component, that is,

$$u(x, 0) = \sum_{n \in \mathbb{N}} \alpha_n e_n, \quad u_t(x, 0) = 0, \tag{3.11}$$

for which (3.10) becomes

$$\sum_{n \neq j} \alpha_n^2 < \eta^4 \alpha_j^2. \tag{3.12}$$

When a fluid hits a bluff body, its flow is modified and goes around the body. Behind the body, due to the lower pressure, the flow creates vortices which appear somehow periodic-in-time and whose frequency depends on the velocity of the fluid. The asymmetry of the vortices generates a forcing lift which starts the vertical oscillations of the structure. The frequencies of the vortices are divided in ranges: for each range of frequencies, there exists a vibrating mode of the structure which captures almost all the input of energy from the vortices [9], giving rise to what we call a solution with a prevailing mode. This is a simplified explanation, further vortices and more complicated phenomena may appear: however, up to some minor details, this explanation is shared by the whole engineering community and it has been studied with some precision in wind tunnel tests, see e.g. [25, 31]. Our analysis starts at this point, that is, when the oscillation is maintained in amplitude by a somehow perfect equilibrium between the input of energy from the wind and the structural dissipation. Whence, as long as the aerodynamics is involved the frequency of the vortices determines the prevailing mode, while in an isolated system the prevailing mode is determined by the initial conditions. This justifies the choice of the initial conditions (3.11).

The behavior of the solutions of (3.1) having a prevailing mode strongly depends on the parameters therein. Two fairly different cases should be distinguished:

$$\gamma_2 = 0 \text{ and } f(s) = \gamma s \quad \text{versus} \quad \gamma_2 \neq 0 \text{ or } f \text{ acts nonlinearly.} \tag{3.13}$$

By f nonlinear, we mean that there exist no neighborhood of $s = 0$ and no $\gamma \geq 0$ for which $f(s) = \gamma s$. In order to emphasize the difference between the two situations in (3.13), let us introduce the notion of invariant space.

Definition 3.3 A subspace $X \subset V(I)$ is called *invariant* for (3.1) under conditions (3.11) if

$$u(x, 0), u_t(x, 0) \in X \implies u(x, t) \in X \text{ for every } t > 0.$$

Of course, $\{0\}$ and $V(I)$ are trivial invariant subspaces. However, in the first case in (3.13) there are many more invariant subspaces than in the second one.

Proposition 3.2 *The following statements hold:*
- *if $\gamma_2 = 0$ and $f(s) = \gamma s$ for some $\gamma \geqslant 0$, then any subspace $X \subset V(I)$ is invariant for (3.1);*
- *if $\gamma_2 \neq 0$ or f is nonlinear, then there are no nontrivial finite-dimensional invariant subspaces for (3.1);*
- *if f is odd, then $X = \langle e_{2n} \rangle_{n \in \mathbb{N}}$ and $X = \langle e_{2n+1} \rangle_{n \in \mathbb{N}}$ are invariant subspaces for (3.1).*

The first statement is straightforward and will be clarified in the course, see Sect. 3.4.1 below. The second statement may also be obtained with some computations, see Sect. 3.5.1 below for the details. The third statement follows by direct computation as well, see [16].

Proposition 3.2 states that the second alternative in (3.13) forces the invariant subspaces for (3.1) to be infinite-dimensional, whereas in the first case there are finite-dimensional invariant subspaces, including one-dimensional subspaces. This striking difference is even more evident if we fix some $j \in \mathbb{N}$ and some $\alpha_j \neq 0$, and we take initial data as in (3.7):

$$u(x, 0) = \alpha_j e_j, \quad u_t(x, 0) = 0. \tag{3.14}$$

Then, in the first case of (3.13), we may simply find the solution of (3.1) with j-th prevailing mode, having the form $u(x, t) = \varphi_j(t) e_j(x)$, for some $\varphi_j \in C^2(\mathbb{R}_+)$. On the contrary, in the second case of (3.13), the same initial conditions give rise to a solution with infinitely many nontrivial Fourier coefficients $\varphi_n(t)$, as soon as $t > 0$.

3.3 Nonlinear Instability and Critical Energy Thresholds

We are interested in the following notion of nonlinear instability, introduced in [16].

Definition 3.4 Let $T_W > 0$. We say that a weak solution $u \in \mathbb{P}_j$ of (3.1) is *unstable before time* $T > 2T_W$ if there exist a residual mode k and a time instant τ with $2T_W < \tau < T$ such that

$$(i) \ \|\varphi_k\|_{L^\infty(0,\tau)} > \eta \|\varphi_j\|_{L^\infty(0,\tau)} \quad \text{and} \quad (ii) \ \frac{\|\varphi_k\|_{L^\infty(0,\tau)}}{\|\varphi_k\|_{L^\infty(0,\tau/2)}} > \frac{1}{\eta} \tag{3.15}$$

(where η is the number appearing in Definition 3.2). We say that u is *stable until time* T if, for any $k \neq j$, (3.15) is not fulfilled for any $\tau \in (2T_W, T)$.

Clearly, this is a "numerical definition", which seems difficult to be characterized analytically with great precision. Roughly speaking, this notion of instability extends in a quantitative way the classical linear instability to a nonlinear context. In all our discussion, we will set $\eta = 0.1$; with this choice, instability is somehow identified with a sufficiently abrupt change in the order of magnitude of the k-th Fourier component of $u \in \mathbb{P}_j$. In Sect. 4.5, we explain in detail how to fulfill it. As pointed out in [16], Definition 3.4 is necessary to

distinguish between physiological energy transfers and instability.

The former occur at any energy level, are unavoidable and start instantaneously, being governed by the natural interactions between modes created by the particular nonlinearity in (3.1). The latter occur and become visible thanks to a (possibly delayed) sudden and violent transfer of energy from some modes to other ones, only occurring at high energies. Condition (i) in (3.15) refers to the significance of the energy transfer, while (ii) refers to its abruptness. The interval $[0, T_W]$ in Definition 3.4 represents a transient phase corresponding to the so-called *Wagner effect* [34], consisting in a time delay in the appearance of the response to a sudden change of the action of an external input in a system. We will determine a reasonable value for T_W in Sect. 3.4.2 (formula (3.40)), after recalling the notion of linear instability. After having fixed $T_W > 0$ and $T > 2T_W$, for all $j \in \mathbb{N}$ the most natural way to define the *j-th energy threshold* of instability before time T for (3.1) would be the following:

$$E_j(a) = \inf_{u \in \mathbb{P}_j} \{\mathcal{E}(u) \mid u \text{ is unstable before time } T\} \qquad \forall a \in (0, 1), \qquad (3.16)$$

where

$$\mathcal{E}(u) = \frac{\|u_t\|_{L^2}^2}{2} + \frac{\|u_{xx}\|_{L^2}^2}{2} + \gamma_1 \frac{\|u_{xx}\|_{L^2}^4}{4} + \gamma_2 \frac{\|u_x\|_{L^2}^4}{4} + \gamma_3 \frac{\|u\|_{L^2}^4}{4} + \int_I F(u)$$

is the constant total energy associated with (3.1) and F is the primitive of f. But then one may not be able to define the *critical energy threshold* of (3.1) as $\inf_j E_j(a)$, since this number could be zero: this fact will be further motivated in Sect. 3.4.3. A possible way out is then to restrict the attention to a finite number of modes, a fairly common procedure in classical engineering literature. Bleich-McCullough-Rosecrans-Vincent [8, p.23] write that *out of the infinite number of possible modes of motion in which a suspension bridge might vibrate, we are interested only in a few, to wit: the ones having the smaller numbers of loops or half waves.* The justification of this approach has physical roots: Smith-Vincent [32, p.11] write that *the higher modes with their shorter waves involve sharper curvature in the truss and, therefore, greater bending moment at a given amplitude and accordingly reflect the influence of the truss stiffness to a greater degree than do the lower modes.* This also occurs in our setting: if $j \to \infty$, this means extremely small oscillations of the prevailing mode. Therefore, the mathematical lack of compactness corresponds to a physically irrelevant phenomenon involving tiny oscillations of high modes, which can thus be neglected.

So, let us fix a finite number of modes. From the Federal Report [1] we learn that, at the TNB, oscillations with more than 10 nodes on the three spans were never seen, see also Fig. 2.2. This means that 12 modes are more than enough to approximate the motion of real bridges and of beams having similar structural responses. From a mathematical point of view, this finite-dimensional approximation is fully justified by the Galerkin procedure that can be used to study (3.1), see [16] for more details; moreover, it enables us to use Proposition 3.1 also in the case $\gamma_1 > 0$. Therefore, we consider the space

$$V_{12}(I) := \mathrm{span}\{e_i;\ i = 0, \ldots, 11\}$$

and we point out that, if the first alternative in (3.13) holds, then $V_{12}(I)$ is invariant, see Proposition 3.2. Then, we seek *approximated solutions* of the projection of equation (3.1) onto $V_{12}(I)$ in the form

$$U^A(x, t) = \sum_{n=0}^{11} \varphi_n(t) e_n(x), \tag{3.17}$$

obtaining the system of 12 nonlinear ODE's ($n = 0, \ldots, 11$)

$$\int_I U^A_{tt} e_n + (1 + \gamma_1 \|U^A_{xx}\|^2_{L^2}) \int_I U^A_{xx} e''_n + \gamma_2 \|U^A_x\|^2_{L^2} \int_I U^A_x e'_n$$

$$+ \gamma_3 \|U^A\|^2_{L^2} \int_I U^A e_n + \int_I f(U^A) e_n = 0. \tag{3.18}$$

These equations are derived from Definition 3.1, by taking U^A and v both in $V_{12}(I)$ and not in the whole space $V(I)$. The function U^A is an approximation of a solution u of (3.1): if instead of just 12 modes we considered an arbitrary number of modes N and we let $N \to \infty$, then it would be $U^A \to u$ in a suitable sense, see [16, Theorem 2].

For each $j = 0, \ldots, 11$, we take initial conditions being the projection of (3.11) on $V_{12}(I)$ satisfying (3.12):

$$U^A(x, 0) = \sum_{n=0}^{11} \alpha_n e_n, \quad U^A_t(x, 0) = 0, \quad \sum_{n \neq j} \alpha^2_n < \eta^4 \alpha^2_j.$$

Then, instead of (3.16), we define

$$E_j(a) = \inf_{U^A \in \mathbb{P}_j} \{\mathcal{E}_j(U^A) \mid U^A \text{ is unstable before time } T\} \quad \forall a \in (0, 1), \quad (3.19)$$

where $\mathcal{E}_j(U^A) := \mathcal{E}(U^A_j)$, U^A_j being the solution of (3.18) having initial data (3.14) (we are thus omitting all the energies involving residual modes; since the residual modes are small, such energies are practically negligible and it is $\mathcal{E}_j(U^A) \approx \mathcal{E}(U^A)$). We are now ready to define what we mean by *energy threshold of instability*.

Definition 3.5 The *energy threshold of instability before time* T for (3.1) is defined as

$$\mathbb{E}_{12}(a) = \inf_{0 \leqslant j \leqslant 11} E_j(a). \tag{3.20}$$

Some remarks are in order. The dependence of E_j on a is due to the boundary-internal conditions (3.4), which modify the spaces $V(I)$ and $V_{12}(I)$ where (3.18) is settled. Our purpose is to detect the "optimal" value for a, for which the threshold $\mathbb{E}_{12}(a)$ is as large as possible, since this value is a measure of the global stability of the approximated beam, truncated on its lowest 12 modes. When (3.18) is governed by energies $\mathcal{E} < \mathbb{E}_{12}(a)$, the oscillations are stable and approximate solutions maintain any prevailing mode before time T, whereas the oscillations of the residual ones remain small and regular. This is no longer true when $\mathcal{E} > \mathbb{E}_{12}(a)$, since there may be a sudden transfer of energy from the prevailing mode to residual modes. If every approximate solution $U^A \in \mathbb{P}_j$ of (3.18) is stable before time T, then $E_j(a) = +\infty$; if this occurs for every j, one has $\mathbb{E}_{12}(a) = +\infty$. In this case, the considered beam is totally stable before time T and nothing else has to be said, otherwise we can prove that the map $a \mapsto \mathbb{E}_{12}(a)$ is continuous.

Theorem 3.1 *Fix $T > 0$ and assume that $\mathbb{E}_{12}(\bar{a}) < +\infty$ for a certain $\bar{a} \in (0, 1)$. Then, $a \mapsto \mathbb{E}_{12}(a)$ is continuous in $\bar{a}$ for (3.18).*

Proof Since we consider variable $a \in (0, 1)$, throughout this proof we emphasize the dependence of the eigenfunctions and eigenvalues of (2.8) on a by denoting them, respectively, by e_n^a and λ_n^a, for $n = 0, \ldots, 11$.

By assumption, $\mathbb{E}_{12}(\bar{a}) < +\infty$ for a certain $\bar{a} \in (0, 1)$. By Definition 3.5, for the choice $a = \bar{a}$ there exist a prevailing mode j, a residual mode $k \neq j$, and a time instant $\tau \in (2T_W, T)$ such that, writing

$$U_{\bar{a}}^A(x, t) := \sum_{n=0}^{11} \varphi_n(t) e_n^{\bar{a}}(x),$$

the conditions in Definitions 3.2 and 3.4 are fulfilled. This means that

$$\sum_{\substack{n=0 \\ n \neq j}}^{11} \left[\varphi_n(0)^2 + \dot{\varphi}_n(0)^2 \right] < \eta^4 (\varphi_j(0)^2 + \dot{\varphi}_j(0)^2) \tag{3.21}$$

and

$$\|\varphi_k\|_{L^\infty(0,\tau)} > \eta \|\varphi_j\|_{L^\infty(0,\tau)}, \quad \frac{\|\varphi_k\|_{L^\infty(0,\tau)}}{\|\varphi_k\|_{L^\infty(0,\tau/2)}} > \frac{1}{\eta}. \tag{3.22}$$

The components φ_j and φ_k solve the two ODEs

$$\ddot{\varphi}_j(t) + \mu_j\varphi_j(t) + \gamma_2\Big(\sum_{\substack{n \bmod 2 \\ =m \bmod 2}} \varphi_n(t)\varphi_m(t)\int_I (e_n^{\bar{a}})'(e_m^{\bar{a}})'\Big)\Big(\sum_{\substack{l \bmod 2 \\ =j \bmod 2}} \varphi_l(t)\int_I (e_l^{\bar{a}})'(e_j^{\bar{a}})'\Big)$$

$$+\gamma_1\Big(\sum_n \mu_n\varphi_n(t)^2\Big)\mu_j\varphi_j(t) + \gamma_3\Big(\sum_n \varphi_n(t)^2\Big)\varphi_j(t) + \int_I f\Big(\sum_n \varphi_n(t)e_n^{\bar{a}}\Big)e_j^{\bar{a}} = 0,$$

and

$$\ddot{\varphi}_k(t) + \mu_k\varphi_k(t) + \gamma_2\Big(\sum_{\substack{n \bmod 2 \\ =m \bmod 2}} \varphi_n(t)\varphi_m(t)\int_I (e_n^{\bar{a}})'(e_m^{\bar{a}})'\Big)\Big(\sum_{\substack{l \bmod 2 \\ =k \bmod 2}} \varphi_l(t)\int_I (e_l^{\bar{a}})'(e_k^{\bar{a}})'\Big)$$

$$\gamma_1\Big(\sum_n \mu_n\varphi_n(t)^2\Big)\mu_k\varphi_k(t) + \gamma_3\Big(\sum_n \varphi_n(t)^2\Big)\varphi_k(t) + \int_I f\Big(\sum_n \varphi_n(t)e_n^{\bar{a}}\Big)e_k^{\bar{a}} = 0.$$

Consider a sequence $\{a_i\}_i$ such that $a_i \to \bar{a}$ for $i \to +\infty$; then,

$$e_n^{a_i} \to e_n^{\bar{a}} \quad \text{in } H^2(I) \quad \text{for all } n = 0, \ldots, 11. \tag{3.23}$$

Indeed, from (2.8) and Theorem 2.4, it follows that

$$\|e_n^{a_i}\|_V^2 = (\lambda_n^{a_i})^4 \to (\lambda_n^{\bar{a}})^4 = \|e_n^{\bar{a}}\|_V,$$

after recalling that all the eigenfunctions are L^2-normalized. This fact shows that there exists $\bar{e} \in V(I)$ such that $e_n^{a_i} \rightharpoonup \bar{e}$ in $H^2(I)$, up to a subsequence. Together with the convergence $e_n^{a_i} \to e_n^{\bar{a}}$ in $L^2(I)$ and the convergence of the norms, this proves (3.23).

For every $i \in \mathbb{N}$, consider now system (3.18) for $a = a_i$ with initial conditions

$$u_i^A(x, 0) = \sum_{n=0}^{11} \varphi_n^i(0)e_n^{a_i}(x), \quad (u_i^A)_t(x, 0) = \sum_{n=0}^{11} \dot{\varphi}_n^i(0)e_n^{a_i}(x),$$

where $\varphi_n^i(0)$ and $\dot{\varphi}_n^i(0)$ are taken so that the associated total energy is equal to $\mathbb{E}_{12}(\bar{a})$ and so that they fulfill (3.21) (this is possible up to taking i sufficiently large, thanks to the strict inequality in (3.21)). Denoting the solution by

$$u_i^A(x, t) = \sum_{n=0}^{11} \varphi_n^i(t)e_n^{a_i}(x),$$

the components φ_j^i and φ_k^i solve the system

$$\ddot{\varphi}_j^i(t) + \mu_j^i\varphi_j^i(t) + \gamma_2\Big(\sum_{\substack{n \bmod 2 \\ =m \bmod 2}} \varphi_n^i(t)\varphi_m^i(t)\int_I (e_n^{a_i})'(e_m^{a_i})'\Big)\Big(\sum_{\substack{l \bmod 2 \\ =j \bmod 2}} \varphi_l^i(t)\int_I (e_l^{a_i})'(e_j^{a_i})'\Big)$$

$$+\gamma_1\Big(\sum_n \mu_n^i\varphi_n^i(t)^2\Big)\mu_j^i\varphi_j^i(t) + \gamma_3\Big(\sum_n \varphi_n^i(t)^2\Big)\varphi_j^i(t) + \int_I f\Big(\sum_n \varphi_n^i(t)e_n^{a_i}\Big)e_j^{a_i} = 0,$$

and

$$\ddot{\varphi}_k^i(t) + \mu_k^i \varphi_k^i(t) + \gamma_2 \Big(\sum_{\substack{n \bmod 2 \\ =m \bmod 2}} \varphi_n^i(t) \varphi_m^i(t) \int_I (e_n^{a_i})'(e_m^{a_i})' \Big) \Big(\sum_{\substack{l \bmod 2 \\ =k \bmod 2}} \varphi_l^i(t) \int_I (e_l^{a_i})'(e_k^{a_i})' \Big)$$

$$+ \gamma_1 \Big(\sum_n \mu_n^i \varphi_n^i(t)^2 \Big) \mu_k^i \varphi_k^i(t) + \gamma_3 \Big(\sum_n \varphi_n^i(t)^2 \Big) \varphi_k^i(t) + \int_I f \Big(\sum_n \varphi_n^i(t) e_n^{a_i} \Big) e_k^{a_i} = 0.$$

From (3.23) and the embedding $H^2(I) \subset C^1(\bar{I})$ we infer the C^1-convergence of $e_n^{a_i}$ to $e_n^{\bar{a}}$, so that also all the integral terms converge. Then, from the Lipschitz continuity of f and the continuous dependence, we infer that $\varphi_n^i \to \varphi_n$ pointwise for every n. Moreover, since all the sequences $\{\varphi_n^i\}_i$ ($n = 0, \dots, 11$) are bounded in H^2, we infer that $\varphi_n^i \to \varphi_n$ in C^1. Hence, there exists i_0 such that, for every $i > i_0$, the components φ_j^i and φ_k^i satisfy (3.22). Since the energy is preserved, we thus have that

$$\mathbb{E}_{12}(\bar{a}) \geqslant \limsup_{i \to +\infty} \mathbb{E}_{12}(a_i),$$

namely $a \mapsto \mathbb{E}_{12}(a)$ is upper-semicontinuous in $a = \bar{a}$.

Assume now by contradiction that $\liminf_{i \to +\infty} \mathbb{E}_{12}(a_i) < \mathbb{E}_{12}(\bar{a})$ and let j_i, k_i be the prevailing and the residual modes fulfilling Definition 3.4 for $a = a_i$. Let us restrict to a subsequence $\{a_{i_l}\}_l$ such that $\lim_{l \to +\infty} \mathbb{E}_{12}(a_{i_l}) = \liminf_{i \to +\infty} \mathbb{E}_{12}(a_i) < \mathbb{E}_{12}(\bar{a}) < +\infty$, so that the sequence $\{\mathbb{E}_{12}(a_{i_l})\}_l$ is bounded. Since the number of considered modes is finite, passing to a further subsequence i_{l_m} it is possible to extract, from the sequence of couples (j_{i_l}, k_{i_l}), a couple of modes (j, k) whose associated Fourier components $(\varphi_j^{i_{l_m}}, \varphi_k^{i_{l_m}})$ fulfill the conditions for instability for every $a = a_{i_{l_m}}$. By continuous dependence, $(\varphi_j^{i_{l_m}}, \varphi_k^{i_{l_m}})$ converges to a couple (φ_j, φ_k) fulfilling the same requirements for $a = \bar{a}$ and the contradiction is reached. This proves lower semicontinuity and completes the proof. $\square$

In our numerical experiments we always found that $\mathbb{E}_{12}(a) < \infty$, which leads to the conjecture that $\mathbb{E}_{12} \in C^0([0, 1])$.

The stability analysis is fairly different for the two cases in (3.13), which will be studied separately. For the first case, Proposition 3.2 states that every subspace of $V(I)$ is invariant, meaning that the modes do not mix or, equivalently, that there is only a low physiological energy transfer. In the second case of (3.13), the modes mix and there are significant physiological energy transfers. Section 3.4 is devoted to the former situation, which is much simpler and will be used to better understand the essence of Definition 3.4 through the notion of *linear stability*. Section 3.5 is instead dedicated to the latter situation.

3.4   Equations That Do Not Mix the Modes

From now on, depending on the context, we denote the eigenfunctions of (2.8) both by e_λ (with associated eigenvalue $\mu = \lambda^4$) and by e_n (with eigenvalue $\mu_n = \lambda_n^4$).

3.4.1   Reduction to a Two-Modes System

We start considering the first case of (3.13), assuming that (3.3) holds with

$$\gamma_1 + \gamma_3 > 0, \quad \gamma_2 = 0 \quad \text{and} \quad f(s) \equiv 0. \tag{3.24}$$

This situation will help us to determine a suitable value for the Wagner time T_W, see (3.40) in Sect. 3.4.2. If we assume (3.24), then (3.1) becomes

$$u_{tt} + u_{xxxx} + \gamma_1 \|u_{xx}\|_{L^2}^2 u_{xxxx} + \gamma_3 \|u\|_{L^2}^2 u = 0, \tag{3.25}$$

to be intended in the weak sense of Definition 3.1. We will treat separately the cases $(\gamma_1, \gamma_3) = (1, 0)$ and $(\gamma_1, \gamma_3) = (0, 1)$ but, at this stage, we deal with any nontrivial couple (γ_1, γ_3) satisfying (3.3). Assuming (3.24), we consider the initial conditions

$$u(x, 0) = \delta e_\lambda, \quad u_t(x, 0) = 0, \tag{3.26}$$

for some $\delta > 0$ and some (L^2-normalized) eigenfunction e_λ of (2.8), relative to the eigenvalue $\mu = \lambda^4$. Then we have the following statement.

Proposition 3.3 *The unique weak solution of* (3.25) *satisfying* (3.26) *has the form*

$$u_\lambda(x, t) = W_\lambda(t)e_\lambda(x), \tag{3.27}$$

where W_λ solves

$$\ddot{W}_\lambda(t) + \lambda^4 W_\lambda(t) + (\gamma_1 \lambda^8 + \gamma_3) W_\lambda(t)^3 = 0, \quad W_\lambda(0) = \delta, \quad \dot{W}_\lambda(0) = 0. \tag{3.28}$$

Proposition 3.3 may be obtained by combining the uniqueness statement in Proposition 3.1 with the orthogonality of the e_λ's, both in L^2 and in V. Together with Proposition 3.2, this result well explains why, under the assumptions (3.24), there is no need to consider approximate solutions such as (3.17): any subspace of $V(I)$ is invariant, in particular the space $V_{12}(I)$.

We call a function u_λ in the form (3.27) a λ-*nonlinear-mode* of (3.25). In fact, for any eigenvalue λ^4 there exist infinitely many λ-nonlinear-modes, one for each value of δ; they are not proportional to each other and they have different periods. Their shape is described by the solution W_λ of (3.28), which depends on δ: for this reason, with an abuse of language, we call W_λ the λ-nonlinear-mode of (3.25) of amplitude δ.

Motivated again by Proposition 3.2, we now consider solutions of (3.25) having two active nonlinear modes. These solutions allow a simple and precise characterization of the stability of one nonlinear mode with respect to another mode. We consider solutions of (3.25) in the form

$$u(x, t) = w(t)e_\lambda(x) + z(t)e_\rho(x) \tag{3.29}$$

for some different eigenfunctions e_λ and e_ρ of (2.8). Notice that, in view of the mutual orthogonality of the eigenfunctions (in L^2 and in V), these solutions appear whenever the initial data (potential and kinetic) are completely concentrated on e_λ and e_ρ, since the vector space generated by e_λ and e_ρ is invariant for the dynamics of (3.25), see Proposition 3.2. After inserting (3.29) into (3.25), we reach the following nonlinear system of ODE's:

$$\begin{cases} \ddot{w}(t) + \lambda^4 w(t) + \Big((\gamma_1\lambda^8 + \gamma_3)w(t)^2 + (\gamma_1\lambda^4\rho^4 + \gamma_3)z(t)^2\Big)w(t) = 0 \\ \ddot{z}(t) + \rho^4 z(t) + \Big((\gamma_1\lambda^4\rho^4 + \gamma_3)w(t)^2 + (\gamma_1\rho^8 + \gamma_3)z(t)^2\Big)z(t) = 0, \end{cases} \tag{3.30}$$

that we associate with the initial conditions

$$w(0) = \delta > 0, \quad z(0) = z_0, \quad \dot{w}(0) = \dot{z}(0) = 0. \tag{3.31}$$

If $z_0 = 0$, then the solution of (3.30)–(3.31) is $(w, z) = (W_\lambda, 0)$ where W_λ is the solution of (3.28), since in this case there is no physiological energy transfer towards modes that are initially zero. In order to characterize the stability of the nonlinear mode W_λ with respect to the nonlinear mode W_ρ we argue as follows. We take initial data in (3.31) such that

$$0 < |z_0| < \eta^2\delta \quad \text{for some } \eta \in (0, 1).$$

This means that the energy of the autonomous system (3.30) is initially almost totally concentrated on the λ-nonlinear-mode, which thus plays the role of the prevailing mode, according to Definition 3.2. One should then wonder whether this remains true for all time $t > 0$ for the solution of (3.30)–(3.31).

★ We consider first the nonlinear equation

$$u_{tt} + \big(1 + \|u_{xx}\|_{L^2}^2\big)u_{xxxx} = 0 \quad x \in I, \quad t > 0, \tag{3.32}$$

which corresponds to (3.25) with $(\gamma_1, \gamma_3) = (1, 0)$. According to Definition 3.1, a weak solution u of (3.32) satisfies $u \in C^0(\mathbb{R}_+; V(I)) \cap C^1(\mathbb{R}_+; L^2(I)) \cap C^2(\mathbb{R}_+; V'(I))$ and

$$\langle u_{tt}, v \rangle_V + \left(1 + \|u_{xx}\|_{L^2}^2\right) \int_I u_{xx} v'' = 0 \qquad \forall v \in V(I), \quad t > 0.$$

Since $(\gamma_1, \gamma_3) = (1, 0)$, problem (3.28) becomes

$$\ddot{W}_\lambda(t) + \lambda^4 W_\lambda(t) + \lambda^8 W_\lambda(t)^3 = 0, \quad W_\lambda(0) = \delta, \quad \dot{W}_\lambda(0) = 0. \qquad (3.33)$$

It is well-known [33] that the unique solution of (3.33) is periodic and that δ is the amplitude of oscillation of W_λ. Moreover, (3.30) becomes

$$\begin{cases} \ddot{w}(t) + \lambda^4 w(t) + \lambda^4 \left(\lambda^4 w(t)^2 + \rho^4 z(t)^2\right) w(t) = 0 \\ \ddot{z}(t) + \rho^4 z(t) + \rho^4 \left(\lambda^4 w(t)^2 + \rho^4 z(t)^2\right) z(t) = 0, \end{cases} \qquad (3.34)$$

which is complemented with the initial conditions (3.31), with $0 < |z_0| < \eta^2 \delta$.

♦ We then consider the nonlinear equation

$$u_{tt} + u_{xxxx} + \|u\|_{L^2}^2 u = 0 \qquad x \in I, \quad t > 0, \qquad (3.35)$$

which corresponds to (3.25) with $(\gamma_1, \gamma_3) = (0, 1)$. According to Definition 3.1, the solution has to be intended in the following weak sense:

$$\langle u_{tt}, v \rangle_V + \int_I u_{xx} v'' + \|u\|_{L^2}^2 \int_I uv = 0 \qquad \forall v \in V(I), \quad t > 0$$

and satisfies $u \in C^0(\mathbb{R}_+; V(I)) \cap C^1(\mathbb{R}_+; L^2(I)) \cap C^2(\mathbb{R}_+; V'(I))$.

Equation (3.35) has a structure similar to (3.32), although the stability analysis gives quite different responses. Since $(\gamma_1, \gamma_3) = (0, 1)$, problem (3.28) becomes

$$\ddot{W}_\lambda(t) + \lambda^4 W_\lambda(t) + W_\lambda(t)^3 = 0, \quad W_\lambda(0) = \delta, \quad \dot{W}_\lambda(0) = 0; \qquad (3.36)$$

compared with (3.33), there is no coefficient λ^8 in front of the cubic term. Moreover, (3.30) becomes the following system of ODE's:

$$\begin{cases} \ddot{w}(t) + \lambda^4 w(t) + \left(w(t)^2 + z(t)^2\right) w(t) = 0 \\ \ddot{z}(t) + \rho^4 z(t) + \left(w(t)^2 + z(t)^2\right) z(t) = 0, \end{cases} \qquad (3.37)$$

to be compared with (3.34). We add to (3.37) the initial conditions (3.31), with $0 < |z_0| < \eta^2 \delta$.

3.4.2 Some Remarks on the Linear Instability

In order to explain the relationships between Definition 3.4 and the classical Floquet theory, we need to recall and discuss the notion of linear instability. This is much simpler when the modes of (3.1) do not mix, for instance under the assumptions (3.24). This discussion is the subject of the next three subsections.

The equation in (3.28) is named after Duffing, due to its first appearance in [14]. If we set

$$b := \sqrt{\lambda^4 + \delta^2(\gamma_1\lambda^8 + \gamma_3)} \quad \text{and} \quad \beta := \frac{\delta}{b}\sqrt{\frac{\gamma_1\lambda^8 + \gamma_3}{2}},$$

then from [11] (see also [33]) we learn that

$$W_\lambda(t) = \delta \operatorname{cn}(bt, \beta), \tag{3.38}$$

where cn is the Jacobi cosine, so that W_λ is periodic with period

$$T(\delta) = \frac{4}{b}\int_0^{\pi/2} \frac{d\varphi}{\sqrt{1 - \beta^2 \sin^2\varphi}}. \tag{3.39}$$

The time-dependent coefficient W_λ in (3.27) starts at level $\delta > 0$ with null first derivative, that is, at a maximum point. Then the number $T(\delta)$ in (3.39) is the time needed for W_λ to complete one cycle and to reach the subsequent maximum point. Whence, if we consider a solution u of (3.1) with an η-prevailing mode (Definition 3.2), the number $T(\delta)$ is a good approximation of the time needed for the Fourier component of the prevailing mode to complete one cycle and hence for the coupling effects between the modes to take place, even in the case where (3.24) does not hold. Therefore, we choose this number as the Wagner time

$$T_W = T(\delta) \tag{3.40}$$

and our numerical experiments confirm that this choice for T_W is good enough to highlight the physiological transfer of energy due to the nonlinearities in (3.1). Clearly, $T_W = T(\delta)$ depends on the amount of energy within (3.1).

The Duffing equation (3.28) is extremely useful to study the *linear instability* of a solution with a prevailing mode.

Definition 3.6 The λ-nonlinear mode W_λ is said to be *linearly stable (unstable)* with respect to the ρ-nonlinear-mode W_ρ if $\xi \equiv 0$ is a stable (unstable) solution of the linear Hill equation

$$\ddot{\xi}(t) + \left(\rho^4 + (\gamma_1\lambda^4\rho^4 + \gamma_3)W_\lambda(t)^2\right)\xi(t) = 0. \tag{3.41}$$

Equation (3.41) comes from the linearization of (3.30) around its solution $(w, z) = (W_\lambda, 0)$. There exist also alternative definitions of stability which, in some cases, can be shown to be equivalent; see e.g. [21]. Moreover, for nonlinear PDE's such as (3.25), Definition 3.6 is appropriate to characterize the instabilities of the nonlinear modes of the equation, see [5, 6, 15].

An important tool to quantify the instability of a general Hill equation of the form

$$\ddot{q}(t) + a(t)q(t) = 0, \qquad a(t + \sigma) = a(t), \tag{3.42}$$

is the *expansion rate* introduced in [2]. The eigenvalues ν_1 and ν_2 of the monodromy matrix associated with (3.42) are the so-called *characteristic multipliers*. The instability for (3.42) takes place when $|\nu_1 + \nu_2| > 2$ (see, e.g., [36]) and the numbers $|\nu_j|^{1/\sigma}$, $j = 1, 2$, represent the growth rates of the amplitude of oscillation for the two solutions of (3.42) having initial values $(q(0), \dot{q}(0)) = (1, 0)$ and $(q(0), \dot{q}(0)) = (0, 1)$, respectively, in the time interval $[0, \sigma]$. Then, following [2], we call *expansion rate* of (3.42) the largest growth rate, namely

$$\mathcal{ER} := \max\{|\nu_1|, |\nu_2|\}^{1/\sigma}$$

and *expansion rate in time* τ the number

$$\mathcal{ER}_\tau := \max\{|\nu_1|, |\nu_2|\}^{\tau/\sigma}, \tag{3.43}$$

which represents the amplitude growth in a lapse of time τ. This will be useful to relate Definitions 3.4 and 3.6 in the forthcoming study of equation (3.25).

★ Let us first deal with (3.34) for which Definition 3.6 may be rephrased by saying that the λ-nonlinear-mode W_λ is linearly stable (unstable) with respect to the ρ-nonlinear-mode W_ρ if $\xi \equiv 0$ is a stable (unstable) solution of the linear Hill equation

$$\ddot{\xi}(t) + \rho^4\left(1 + \lambda^4 W_\lambda(t)^2\right)\xi(t) = 0. \tag{3.44}$$

For all $\delta, \lambda, \rho > 0$, we introduce the parameters

$$\gamma := \frac{\rho^2}{\lambda^2}, \qquad \Lambda_\delta := 2\sqrt{2}\int_0^{\pi/2}\sqrt{\frac{1 + \delta^2\lambda^4\sin^2\theta}{2 + \delta^2\lambda^4 + \delta^2\lambda^4\sin^2\theta}}\,d\theta, \tag{3.45}$$

that play a crucial role in the linear stability of the λ-nonlinear mode with respect to the ρ-nonlinear mode. Let us also introduce the two sets

$$I_S := \bigcup_{k=0}^{+\infty}\left(k(2k + 1), (k + 1)(2k + 1)\right) \qquad I_U := \bigcup_{k=0}^{+\infty}\left((k + 1)(2k + 1), (k + 1)(2k + 3)\right); \tag{3.46}$$

note that $\overline{I_S \cup I_U} = [0, \infty)$. The next statement gives sufficient conditions for the stability/instability of (3.34).

Theorem 3.2 *Let $\lambda^4 \neq \rho^4$ be two eigenvalues of (2.8), let Λ_δ be as in (3.45) and I_S, I_U as in (3.46).*
(i) The λ-nonlinear-mode of (3.32) of amplitude δ is linearly stable with respect to the ρ-nonlinear-mode whenever one of the following facts holds:

$(i)_1$ $\lambda > \rho$ and $\delta > 0$;
$(i)_2$ $\lambda < \rho$ and δ is sufficiently small;
$(i)_3$ $\frac{\rho^4}{\lambda^4} \in I_S$ and δ is sufficiently large;
$(i)_4$ there exists $k \in \mathbb{N}$ such that

$$\log(1 + \delta^2\lambda^4) < 2 \cdot \min\left\{\frac{\rho^2}{\lambda^2}\Lambda_\delta - k\pi, \ (k+1)\pi - \frac{\rho^2}{\lambda^2}\Lambda_\delta\right\}.$$

(ii) The λ-nonlinear-mode of (3.32) of amplitude δ is linearly unstable with respect to the ρ-nonlinear-mode whenever one of the following facts holds:

$(ii)_1$ $1 < \frac{\rho^2}{\lambda^2} < \psi_\lambda(\delta)$, where $\psi_\lambda : \mathbb{R}_+ \to \mathbb{R}_+$ is a continuous function such that $\psi_\lambda(\delta) > 1$ for every $\delta > 0$ and

$$\psi_\lambda(\delta) \leqslant 1 + \left(\frac{1}{8} + \frac{1}{2\pi}\right)\delta^2\lambda^4 + O(\delta^4) \text{ as } \delta \to 0, \qquad \lim_{\delta \to \infty} \psi_\lambda(\delta) = \sqrt{3};$$

$(ii)_2$ $\frac{\rho^4}{\lambda^4} \in I_U$ and δ is sufficiently large.

Proof Since the proof may be obtained by combining several arguments from [18], we only briefly sketch it. We set

$$\Phi_\lambda(t) := \lambda^2 W_\lambda\left(\frac{t}{\lambda^2}\right)$$

and we notice that if W_λ satisfies (3.33), then Φ_λ solves

$$\ddot{\Phi}_\lambda(t) + \Phi_\lambda(t) + \Phi_\lambda(t)^3 = 0, \qquad \Phi_\lambda(0) = \lambda^2\delta, \qquad \dot{\Phi}_\lambda(0) = 0. \qquad (3.47)$$

With these transformations, also (3.44) changes and we see that the λ-nonlinear-mode of (3.32) of amplitude δ is linearly stable with respect to the ρ-nonlinear-mode W_ρ if and only if $\xi \equiv 0$ is a stable solution of

$$\ddot{\xi}(t) + \gamma^2\left(1 + \Phi_\lambda(t)^2\right)\xi(t) = 0. \qquad (3.48)$$

Then we recall a criterion due to Burdina [10] (see also [36, Test 3, Sect. 3, Chap. VIII]), which yields a sufficient condition for the stability of some Hill equations.

Lemma 3.1 *Let $T > 0$ and let p be a continuous, T-periodic and strictly positive function having a unique maximum point and a unique minimum point in $[0, T)$. If there exists $k \in \mathbb{N}$ such that*

$$k\pi + \frac{1}{2}\log\frac{\max p}{\min p} < \int_0^T \sqrt{p(t)}\, dt < (k+1)\pi - \frac{1}{2}\log\frac{\max p}{\min p}, \qquad (3.49)$$

then the trivial solution of the Hill equation $\ddot{\xi}(t) + p(t)\xi(t) = 0$ is stable.

We apply Lemma 3.1 to the Hill equation (3.48), that is, with $p(t) = \gamma^2\big(1 + \Phi_\lambda(t)^2\big)$
so that
$$\log \frac{\max p}{\min p} = \log\big(1 + \delta^2\lambda^4\big).$$

Indeed, from Burgreen [11] we know that

$$\Phi_\lambda(t) = \delta\lambda^2 \operatorname{cn}\left[t\,\sqrt{1 + \delta^2\lambda^4}, \; \frac{\delta\lambda^2}{\sqrt{2(1 + \delta^2\lambda^4)}}\right],$$

where cn is the Jacobi cosine. The period of Φ_λ can be computed via (3.39), so that
the period of Φ_λ^2 is given by

$$T_\lambda(\delta) = 2\sqrt{2} \int_0^{\pi/2} \frac{d\phi}{\sqrt{2 + \delta^2\lambda^4(1 + \sin^2\phi)}} \tag{3.50}$$

and $\lim_{\delta \to 0} T_\lambda(\delta) = \pi$.

Moreover, from the energy conservation in (3.47) we see that

$$2\dot{\Phi}_\lambda(t)^2 = \big(\delta^2\lambda^4 - \Phi_\lambda(t)^2\big)\big(2 + \delta^2\lambda^4 + \Phi_\lambda(t)^2\big);$$

therefore, thanks to symmetries and with the change of variable $\Phi_\lambda(t) = \delta\lambda^2 \sin\theta$,
we obtain

$$\int_0^{T_\lambda(\delta)} \sqrt{1 + \Phi_\lambda(t)^2}\,dt = 2\int_0^{T_\lambda(\delta)/2} \sqrt{1 + \Phi_\lambda(t)^2}\,dt = \Lambda_\delta,$$

see [18] for the details; here, Λ_δ is as in (3.45). Item $(i)_4$ then follows by applying
Lemma 3.1 to (3.48).

By using the very same method as in [12, Theorem 3.1] (see also [5, Theorem
8]), we may obtain the following statement.

Lemma 3.2 *Let I_S and I_U as in (3.46). For every $\gamma > 0$ there exists $\bar{\delta}_\gamma > 0$ such
that, for all $\delta > \bar{\delta}_\gamma$:*
(a) if $\gamma^2 \in I_U$, then the trivial solution of equation (3.48) is unstable;
(b) if $\gamma^2 \in I_S$, then the trivial solution of equation (3.48) is stable.

Lemma 3.2 proves Items $(i)_3$ and $(ii)_2$ of Theorem 3.2.

When $\gamma = 1$ we see that $\xi = \Phi_\lambda$ solves (3.48); even if the case $\lambda = \rho$ has no
meaning for (3.34), this tells us that $\gamma = 1$ is part of the boundary of the resonance
tongue U_1 emanating from the point $(\delta\lambda^2, \gamma) = (0, 1)$: this proves Item $(i)_1$. Moreover, combined with Lemma 3.2 and with the asymptotic behavior

$$\Lambda_\delta = \pi - \frac{\pi}{8}\delta^2\lambda^4 + O(\delta^4) \qquad \text{as } \delta \to 0,$$

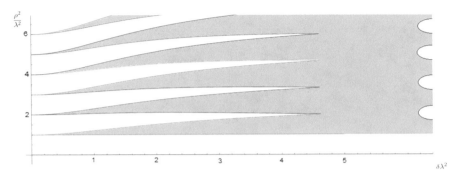

Fig. 3.1 Stability regions (white) obtained with the sufficient conditions (i) of Theorem 3.2

Items $(i)_1$ and $(i)_4$ show that the resonant tongue U_1 of (3.44) emanating from $(\delta\lambda^2, \gamma) = (0, 1)$ is given by

$$U_1 = \left\{(\delta\lambda^2, \gamma) \in \mathbb{R}_+^2 : 1 < \gamma < \psi_\lambda(\delta)\right\}$$

where ψ_λ satisfies the properties stated in Item $(ii)_1$.

Finally, note that Item $(i)_4$ implies that if $k \in \mathbb{N}$ $(k \geqslant 2)$ and if $(\delta\lambda^2, \gamma) \in U_k$, where U_k is the resonant tongue of (3.44) emanating from $(\delta\lambda^2, \gamma) = (0, k)$, then necessarily

$$k + \left(\frac{k}{8} - \frac{1}{2\pi}\right)\delta^2\lambda^4 + O(\delta^4) \leqslant \frac{\rho^2}{\lambda^2} \leqslant k + \left(\frac{k}{8} + \frac{1}{2\pi}\right)\delta^2\lambda^4 + O(\delta^4) \quad \text{as } \delta \to 0.$$
$$(3.51)$$

This proves Item $(i)_2$ and completes the proof of Theorem 3.2. □

From [36, Chap. VIII] we know that for all $k = 1, 2, 3 \ldots$ there exists a resonant tongue U_k emanating from the point $(\delta\lambda^2, \gamma) = (0, k)$. If $(\delta\lambda^2, \gamma)$ belongs to one of these tongues, then the trivial solution of (3.44) is unstable. Since $\Lambda_\delta = \pi(1 - \delta^2\lambda^4/8) + O(\delta^4)$ as $\delta \to 0$, Item $(i)_4$ implies that if $(\delta\lambda^2, \gamma) \in U_k$, then necessarily (3.51) holds. This enables us to depict the resonance tongues in a neighborhood of $\delta = 0$. In Fig. 3.1 we display the (white) regions of stability described by Theorem 3.2-(i).

We also numerically obtained a full picture of the resonance tongues of (3.44). For each value of $\delta = W_\lambda(0)$ we computed the period of W_λ through formula (3.39) and we determined the monodromy matrix of (3.44). Then we computed the trace of the monodromy matrix after one period of time, seeking when it was equal to ± 2. The results are reported in Fig. 3.2. It turns out that the resonance tongues are very narrow for small δ and they enlarge as $\delta \to \infty$. As a consequence of Theorem 3.2 (see also Fig. 3.2) we see that for all $k \in \mathbb{N}$ the resonance lines emanating from the points $(0, k)$ of the $(\delta\lambda^2, \rho^2/\lambda^2)$-plane tend towards $\sqrt{k(2k - 1)}$ and $\sqrt{k(2k + 1)}$ as $\delta \to \infty$. Therefore, the amplitude at infinity of the k-th resonant tongue is given by

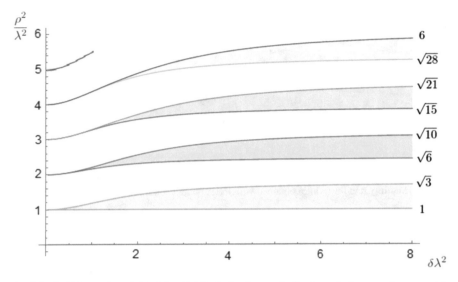

Fig. 3.2 Stability regions (white) for (3.44) obtained numerically through the monodromy matrix

$$\sqrt{k(2k+1)} - \sqrt{k(2k-1)} = \frac{2}{\sqrt{2 + \frac{1}{k}} + \sqrt{2 - \frac{1}{k}}} \in (\sqrt{2}/2, \sqrt{3} - 1] \approx (0.707, 0.732]$$

for every $k \in \mathbb{N}$, which shows that the resonant tongues all have, approximately, the same width. A similar bound holds for the width of the stability regions as $\delta \to \infty$. Once the value of $\gamma = \rho^2/\lambda^2$ is fixed, if δ increases starting from the point $(0, \gamma)$, some narrow resonance tongues are crossed before reaching the final stability or instability region characterized by the sets I_S and I_U.

◆ Let us now turn to the case of (3.37). Following Definition 3.6, we say that the λ-nonlinear-mode W_λ is linearly stable (unstable) with respect to the ρ-nonlinear-mode W_ρ if $\xi \equiv 0$ is a stable (unstable) solution of the linear Hill equation

$$\ddot{\xi}(t) + \left(\rho^4 + W_\lambda(t)^2\right)\xi(t) = 0. \tag{3.52}$$

If W_λ solves (3.36) and we set $W_\lambda(t) = \lambda^2 \Psi_\lambda(\lambda^2 t)$, we see that Ψ_λ satisfies

$$\ddot{\Psi}_\lambda(t) + \Psi_\lambda(t) + \Psi_\lambda(t)^3 = 0, \qquad \Psi_\lambda(0) = \frac{\delta}{\lambda^2}, \qquad \dot{\Psi}_\lambda(0) = 0.$$

The same change of variables shows that the stability of (3.52) is equivalent to the stability of

$$\ddot{\xi}(t) + \left(\frac{\rho^4}{\lambda^4} + \Psi_\lambda(t)^2\right)\xi(t) = 0.$$

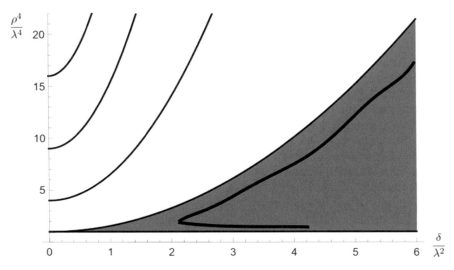

Fig. 3.3 Regions of linear stability (white) and instability (gray) for (3.37)

Then we can take advantage of the results in [18, Sect. 2] and obtain the counterpart of Theorem 3.2.

Proposition 3.4 *Let[1] $\lambda^4 \neq \rho^4$ be two eigenvalues of (2.8). The λ-nonlinear-mode of (3.35) of amplitude δ is linearly stable with respect to the ρ-nonlinear-mode if and only if one of the following facts holds:*

$$\lambda > \rho \ and \ \delta > 0 \quad or \quad \lambda < \rho \ and \ \delta \leqslant \sqrt{2(\rho^4 - \lambda^4)}.$$

Obviously, this means that the λ-nonlinear-mode of (3.35) of amplitude δ is linearly unstable with respect to the ρ-nonlinear-mode if and only if

$$\lambda < \rho \quad and \quad \delta > \sqrt{2(\rho^4 - \lambda^4)}.$$

Hence, Proposition 3.4 states that

> **whenever $\lambda < \rho$, the λ-nonlinear-mode of (3.35) is linearly unstable**
> **with respect to the ρ-nonlinear-mode if and only if**
> **it oscillates with sufficiently large amplitude.**

We may give a full description of the stability and instability regions for (3.37) (see Fig. 3.3, which should be compared with Fig. 3.2).

The lowest parabola in Fig. 3.3 bounds the first instability tongue and has equation

---

[1] The authors are grateful to Clelia Marchionna for raising their attention on the papers [20, 22] which led to an improvement of Proposition 3.4.

$$\frac{\rho^4}{\lambda^4} = 1 + \frac{1}{2} \left(\frac{\delta}{\lambda^2} \right)^2,$$

see [18, Theorem 4]. By using the so-called KdV hierarchy, one finds that the insta-
bility tongues emanating from $(\delta/\lambda^2, \rho^4/\lambda^4) = (0, n^2)$ (with $n \geqslant 2$) are degenerate,
see [20] and also the introduction in [22]. This means that, contrary to Fig. 3.2 and
except for the first tongue, there are no open instability tongues since their upper
and lower boundaries coincide. We determined these degenerate tongues numerically, representing them by the black parabolas in Fig. 3.3. By applying the Burdina
criterion [10] (see also Lemma 3.1) and arguing as for Theorem 3.2, we infer that
the degenerate instability tongues emanating from $(\delta/\lambda^2, \rho^4/\lambda^4) = (0, n^2)$ $(n \geqslant 2)$
asymptotically satisfy, for $\delta \to 0$, the inequalities

$$n^2 + \left(\frac{3n^2}{4} - \frac{1}{2} - \frac{1}{\pi n} \right) \frac{\delta^2}{\lambda^4} + O(\delta^4) \leqslant \frac{\rho^4}{\lambda^4} \leqslant n^2 + \left(\frac{3n^2}{4} - \frac{1}{2} + \frac{1}{\pi n} \right) \frac{\delta^2}{\lambda^4} + O(\delta^4).$$

The characterization of the black line within the gray instability region will be given
in Sect. 3.4.5: it represents the thresholds of nonlinear instability.

3.4.3 The Critical Energy Threshold for the Linear Instability

In this section, we compute the critical energy threshold for the linear stability of
Eqs. (3.32) and (3.35). To this end, it will be useful to preliminarily introduce some
notations. For all eigenvalues λ^4 and ρ^4 of (2.8), we define

$$D(\lambda, \rho) := \inf\{d > 0; \ (3.41) \text{ has unstable solutions when } W_\lambda(0) = \delta > d\}.$$
(3.53)

We immediately observe that this quantity may be infinite: for (3.32) and (3.35)
respectively, we have

$$D(\lambda, \rho) \begin{cases} = +\infty & \text{if } \rho^4/\lambda^4 \in I_S \\ < +\infty & \text{if } \rho^4/\lambda^4 \in I_U \end{cases} \quad \text{and} \quad D(\lambda, \rho) = \begin{cases} +\infty & \text{if } \rho < \lambda \\ \sqrt{2(\rho^4 - \lambda^4)} & \text{if } \rho > \lambda \end{cases}$$

in view of Theorem 3.2 (Fig. 3.2) for (3.32) and of Proposition 3.4 (Fig. 3.3) for
(3.35).

Then we define the critical energy as the (constant) energy of the solutions of
(3.33) and (3.36), respectively, when $\delta = D(\lambda, \rho)$:

$$E(\lambda, \rho) := \frac{\lambda^4 D(\lambda, \rho)^2}{2} + \frac{\lambda^8 D(\lambda, \rho)^4}{4}, \qquad E(\lambda, \rho) := \frac{\lambda^4 D(\lambda, \rho)^2}{2} + \frac{D(\lambda, \rho)^4}{4},$$
(3.54)

for (3.32) and (3.35), respectively. Also this energy can be infinite: we have, respectively,

$$E(\lambda, \rho) \begin{cases} = +\infty \text{ if } \rho^4/\lambda^4 \in I_S \\ < +\infty \text{ if } \rho^4/\lambda^4 \in I_U \end{cases} \quad \text{and} \quad E(\lambda, \rho) = \begin{cases} +\infty & \text{if } \rho < \lambda \\ (\rho^4 - \lambda^4)\rho^4 & \text{if } \rho > \lambda. \end{cases}$$
$$(3.55)$$

Clearly, the stability of the mode W_λ with respect to W_ρ depends on a through $\lambda = \lambda(a)$ and $\rho = \rho(a)$. For both problems (3.32) and (3.35), the most natural way to define the energy threshold for linear stability in dependence of a would be

$$\mathbb{E}^\ell(a) := \inf_{\lambda, \rho} E\big(\lambda(a), \rho(a)\big);$$

unfortunately, this function has several setbacks. First of all, from (2.11) and Theorem 3.2-$(ii)_1$ we see that, in case (3.32), one has

$$\lim_{n \to \infty} E\big(\lambda_n(a), \lambda_{n+1}(a)\big) = 0 \quad \forall a \in (0, 1),$$

which shows that $\mathbb{E}^\ell(a) \equiv 0$ and thus this threshold is not meaningful. Moreover, as for (3.35), formula (3.55) implies that $\mathbb{E}^\ell(a) < \infty$ for all $a \in (0, 1)$; on the other hand, the distribution of the eigenvalues for $a = 1/2$, described in Corollary 2.1, guarantees that $\mathbb{E}^\ell(1/2) > 0$. A definition of critical energy that has so different behaviors for quite similar nonlinearities does not appear appropriate. Therefore, we are again led to follow Definition 3.5 and set

$$\mathbb{E}^\ell_{12}(a) := \inf_{\lambda_0 \leqslant \lambda < \rho \leqslant \lambda_{11}} E\big(\lambda(a), \rho(a)\big) \tag{3.56}$$

where the constraint $\lambda(a) < \rho(a)$ is due to the fact that for $\lambda(a) > \rho(a)$ there is no instability. This means that there are 66 couples to evaluate and the above infimum is certainly a minimum if at least one of these couples yields finite $E(\lambda(a), \rho(a))$.

In order to determine the position $a \in (0, 1)$ of the piers which maximizes $\mathbb{E}^\ell_{12}(a)$ (yielding more stability), one is then interested in showing that the map $a \mapsto \mathbb{E}^\ell_{12}(a)$ is continuous. This can be easily proved for (3.35).

Proposition 3.5 *For Eq.*(3.35) *the map* $a \mapsto \mathbb{E}^\ell_{12}(a)$ *is strictly positive and continuous.*

Proposition 3.5 follows by combining the continuity of the maps $a \mapsto \lambda(a)$ (see Theorem 2.4) with the explicit form of the energy in (3.55) (second formula): since $\mathbb{E}^\ell_{12}(a)$ is the minimum of a finite number of positive and continuous functions, it is positive and continuous, as well.

However, the map $a \mapsto \mathbb{E}^\ell_{12}(a)$ for (3.32) may not be continuous, as we now explain through a graphic counterexample. Let us take the "inverse" of the plot in Fig. 3.2, namely its reflection about the first bisectrix and let us only plot the four upper boundaries of the resonant tongues, see Fig. 3.4. The black lines represent the value of $\lambda^2 D(\lambda, \rho)$ and, since $E(\lambda, \rho)$ increasingly depends on this value, they maintain the order of $E(\lambda, \rho)$ for different couples (λ, ρ). If $\rho^4/\lambda^4 \in I_U$, we plot the horizontal segments at the top of the picture, which means that $\lambda^2 D(\lambda, \rho) = +\infty$

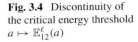

Fig. 3.4 Discontinuity of
the critical energy threshold
$a \mapsto \mathbb{E}_{12}^{\ell}(a)$

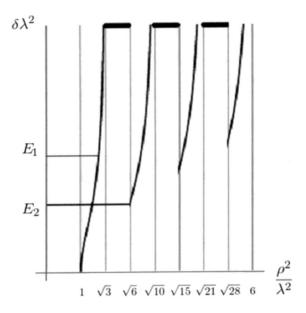

and, in turn, also that $E(\lambda, \rho) = +\infty$. Consider now two couples $(\lambda_i(a), \rho_i(a))$
and $(\lambda_j(a), \rho_j(a))$. From Theorem 2.4 we know that the two ratios $\rho_k(a)/\lambda_k(a)$
(for $k = i, j$) depend continuously on a. Assume that $[\rho_i(a)/\lambda_i(a)]^2$ varies in a left
neighborhood of $\sqrt{3}$ in such a way that $\lambda_i^2 D(\lambda_i, \rho_i)$ varies in a small neighborhood of
E_1 on the vertical axis. Assume also that $[\rho_j(a)/\lambda_j(a)]^2$ varies in an interval centered
at $\sqrt{6}$: then $\lambda_j^2 D(\lambda_j, \rho_j)$ can be either $+\infty$ (in a left neighborhood of $\sqrt{6}$) or around
E_2 (in a right neighborhood of $\sqrt{6}$). The minimum between the two $\lambda_k^2 D(\lambda_k, \rho_k)$
will then be close to E_1 in the first case and close to E_2 in the second case, thereby
displaying discontinuity.

Clearly, this example is based on the fact that the ratios $[\rho(a)/\lambda(a)]^4$ may enter
and exit from I_U. One could still try to prove some continuity provided one could
show that all the involved ratios remain in the same stability/instability interval, but
this seems out of reach. Anyway, according to Fig. 3.2,

$$\textbf{the resonant lines are increasing functions of } \delta\lambda^2. \qquad (3.57)$$

Proving this statement seems to be a difficult task. But, taking it for granted, it enables
to prove the lower semicontinuity of the energy threshold.

Theorem 3.3 *Consider problem* (3.32). *For every* $a \in (0, 1)$ *we have* $0 < \mathbb{E}_{12}^{\ell}(a) <$
$+\infty$. *Moreover, if* (3.57) *holds, then the map* $a \mapsto \mathbb{E}_{12}^{\ell}(a)$ *is lower semicontinuous.*

Proof To prove the first part of the statement, it suffices to show that, for every
$a \in (0, 1)$, at least one of the admissible ratios $\rho^4(a)/\lambda^4(a)$ for the computation of

$\mathbb{E}_{12}^{\ell}(a)$ belongs to the first instability interval $(1, 3)$ of I_U. If $\lambda_{11}^2(a)/\lambda_{10}^2(a) < \sqrt{3}$, we are done. Otherwise, $\lambda_{10}^2(a) \leqslant \lambda_{11}^2(a)/\sqrt{3}$ and from (2.12) we know that

$$5 \leqslant \lambda_9(a) \leqslant 6 \leqslant \lambda_{11}(a) \leqslant 7,$$

implying $\lambda_{10}^2(a) \leqslant 49/\sqrt{3}$ and, in turn,

$$\frac{\lambda_{10}^2(a)}{\lambda_9^2(a)} \leqslant \frac{49/\sqrt{3}}{25} < \sqrt{3}.$$

Therefore, $0 < \mathbb{E}_{12}^{\ell}(a) < +\infty$ for every $a \in (0, 1)$.

For the proof of the lower semicontinuity, we start by noticing that the couples of eigenvalues of (2.8) candidates to minimize E, defined in (3.54), have to be sought among consecutive eigenvalues. So, fix some $\overline{a} \in (0, 1)$ and a couple of consecutive eigenvalues $\lambda_n(a)$ and $\lambda_{n+1}(a)$ with $a \to \overline{a}$. Three cases may occur.

• First case: $\lambda_{n+1}^4(\overline{a})/\lambda_n^4(\overline{a}) \in I_S$ (open interval).

Then $\lambda_{n+1}^4(a)/\lambda_n^4(a) \in I_S$ for a sufficiently close to $\overline{a}$ in view of Theorem 2.4. Next, we need to use tools from the classical Floquet theory, see e.g. [36]. Following Definition 3.6, the values of $D(\lambda_n(a), \lambda_{n+1}(a))$ and $E(\lambda_n(a), \lambda_{n+1}(a))$ depend on the stability of (3.48). So, consider the two solutions ξ_1 and ξ_2 of (3.48) satisfying, respectively, the initial conditions

$$\xi_1(0) = 1, \quad \dot{\xi}_1(0) = 0, \qquad \xi_2(0) = 0, \quad \dot{\xi}_2(0) = 1.$$

From (3.50) we know that the period of Φ_λ^2 is equal to $T_\lambda(\delta)$, so that $\delta \mapsto T_\lambda(\delta)$ is a continuous function. Then, the trace of the monodromy matrix associated with (3.48) is given by

$$\xi_1(T_\lambda(\delta)) + \dot{\xi}_2(T_\lambda(\delta)),$$

hence $D(\lambda_n(a), \lambda_{n+1}(a))$ in (3.53) has the following equivalent characterization:

$$D(\lambda_n(a), \lambda_{n+1}(a)) = \inf \left\{ d > 0; \left| \xi_1(T_\lambda(\delta)) + \dot{\xi}_2(T_\lambda(\delta)) \right| > 2 \text{ when } W_{\lambda_n}(0) = \delta > d \right\}.$$

Since the resonant lines are continuous [36], we have that $D(\lambda_n(a), \lambda_{n+1}(a)) \to D(\lambda_n(\overline{a}), \lambda_{n+1}(\overline{a}))$. By definition of E, this shows that the map $a \mapsto E(\lambda_n(a), \lambda_{n+1}(a))$ is continuous in all the values of a for which $\lambda_{n+1}^4(a)/\lambda_n^4(a) \in I_S$.

• Second case: $\lambda_{n+1}^4(\overline{a})/\lambda_n^4(\overline{a}) \in I_U$ (open interval).

This case is simple, since $E(\lambda_n(a), \lambda_{n+1}(a)) \equiv \infty$ in a neighborhood of $\overline{a}$, which means that the couple $(\lambda_n(a), \lambda_{n+1}(a))$ does not compete to achieve the minimum in E.

• Third case: $\lambda_{n+1}^4(\overline{a})/\lambda_n^4(\overline{a}) \notin I_S \cup I_U$.

From (3.46) we know that there exists an integer $k \geqslant 2$ such that $\lambda_{n+1}^4(\overline{a})/\lambda_n^4(\overline{a}) = k(k + 1)/2$. By Theorem 2.4 we then infer that

$$\lim_{a \to \overline{a}} \frac{\lambda_{n+1}^4(a)}{\lambda_n^4(a)} = \frac{k(k+1)}{2}.$$

Whence, from the continuity of the resonant lines [36] and using (3.57), we have that

$$\liminf_{a \to \overline{a}} E(\lambda_n(a), \lambda_{n+1}(a)) = E(\lambda_n(\overline{a}), \lambda_{n+1}(\overline{a})).$$

This shows that the map $a \mapsto E(\lambda_n(a), \lambda_{n+1}(a))$ is lower semicontinuous in this third case. Incidentally, we observe that (3.57) is needed precisely for the third case.

By combining the analysis for the three cases, we infer that

all the maps $a \mapsto E(\lambda_n(a), \lambda_{n+1}(a))$ are lower semicontinuous in every $\overline{a} \in (0, 1)$.

Since $\mathbb{E}_{12}^{\ell}(a)$ is the minimum of a finite number of lower semicontinuous functions, it is itself lower semicontinuous. $\qquad \square$

3.4.4 Optimal Position of the Piers for the Linear Instability

In this section, we aim at finding the position of the piers which yields more linear stability, for each of the problems (3.32) and (3.35). In other words, we seek the optimal $a \in (0, 1)$ which maximizes $\mathbb{E}_{12}^{\ell}(a)$. As in the previous subsection, the two equations behave quite differently and we analyze them separately.

★ Overall, Theorem 3.2 states that for (3.32) the relevant parameter for stability is γ defined in (3.45), that is, the square root of the ratio of the eigenvalues of (2.8). In particular, instability arises if $\gamma^2(a) \in I_U$ (we emphasize here the dependence on a). In order to study the maps $a \mapsto \gamma(a)$, we take advantage of the results in Chap. 2, in particular of the numerical computation of the eigenvalues. The ratios turn out to be extremely irregular, with no obvious rule governing their variation. In Fig. 3.5 we plot some of the graphs obtained by interpolation after computing the ratios with step 0.05 for a. The shaded horizontal strips correspond to the square roots of the (endpoints of the) intervals composing I_U, see (3.46) and Fig. 3.2. If the value of $\gamma = \lambda_k^2/\lambda_j^2$ is within this range, then for large energies the λ_j-nonlinear-mode W_{λ_j} is linearly unstable with respect to the λ_k-nonlinear-mode W_{λ_k}, otherwise it is linearly stable.

The ratio itself may enter and exit I_U several times, as a varies in $(0, 1)$. We see that λ_1^2/λ_0^2 lies within the stability region for large energies only if (approximately) $0.4 < a < 0.5$. The ratio λ_4^2/λ_0^2 is much more complicated, as a varies it enters and exits several times the instability regions: similar behaviors are visible whenever the ratio between the eigenvalues has large variations for varying a. The ratio $\lambda_{11}^2/\lambda_7^2$ is almost always within the stability region. On the contrary, there exist some ratios which never exit the instability region, see Fig. 3.6: this means that no choice of $a \in$

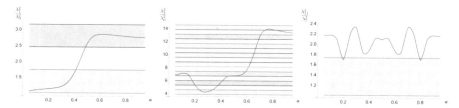

Fig. 3.5 Plot of the maps $a \mapsto \lambda_1^2/\lambda_0^2$, $a \mapsto \lambda_4^2/\lambda_0^2$, $a \mapsto \lambda_{11}^2/\lambda_7^2$ (from left to right) and their intersection with the (shaded) instability intervals

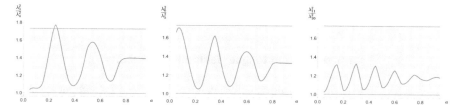

Fig. 3.6 Plot of the maps $a \mapsto \lambda_5^2(a)/\lambda_4^2(a)$, $a \mapsto \lambda_6^2(a)/\lambda_5^2(a)$, $a \mapsto \lambda_{11}^2(a)/\lambda_{10}^2(a)$ (from left to right) and their intersection with the first (shaded) instability interval

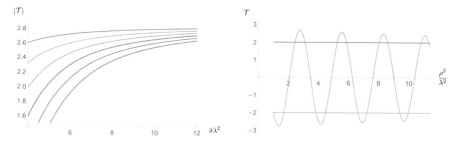

Fig. 3.7 Comparison between the traces of monodromy matrices in different resonant tongues

$(0, 1)$ can prevent the linear instability for these couples of eigenvalues. Therefore, it is difficult to derive a precise rule telling which values of a yield more stability.

Moreover, not all the intervals in I_U yield *quantitatively* the same amount of instability. We order them by means of their instability effects as follows. We compute the absolute value $|\mathcal{T}|$ of the trace of the monodromy matrix associated with (3.44) when $\gamma = \rho^2/\lambda^2 = \frac{\sqrt{k(2k+1)}+\sqrt{k(2k-1)}}{2}$, namely at the midpoint of the limit interval $[\sqrt{k(2k-1)}, \sqrt{k(2k+1)}]$ (as $\delta \to \infty$) of a resonant tongue. We start with $\delta\lambda^2 = 4$ and we proceed by increasing $\delta\lambda^2$ with step 0.2 until $\delta\lambda^2 = 12$, then we interpolate. We do this for several different tongues. In the left picture of Fig. 3.7 we display the plots for the lowest six tongues. It appears clearly that the values of $|\mathcal{T}|$ are ordered, the largest one being the first, and then they decrease as the extremal values of the resonant tongue increase.

This provides the following principle:

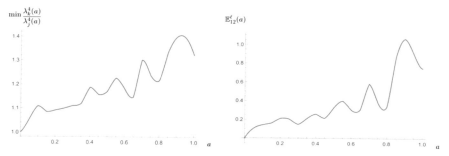

Fig. 3.8 Graph of the map (3.58) and of $a \mapsto \mathbb{E}^{\ell}_{12}(a)$ for (3.32)

the first resonant tongue yields more instability.

We made further experiments, studying the behavior of the trace $\mathcal{T}$ (without absolute value) for fixed $\delta\lambda^2$ on varying of ρ^2: starting from $\rho^2 = 1$ we proceeded with step 0.1 and then we interpolated. In the right picture of Fig. 3.7 we display the plot for $\delta\lambda^2 = 8$. It turns out that the sequence of the maxima of $|\mathcal{T}|$ is decreasing. Other values of this parameter showed the same behavior: $\mathcal{T}$ oscillates above 2 and below -2 with decreasing amplitude. This confirms the above principle.

By the first formula in (3.54), the critical energy threshold depends increasingly on $\delta\lambda^2$. Moreover, the critical amplitude $D(\lambda, \rho)\lambda^2$ depends increasingly on ρ/λ, thus $\mathbb{E}^{\ell}_{12}(a)$ takes its maximum and its minimum at the very same points as γ: the instability increases as γ approaches 1. In the left picture of Fig. 3.8 we display the graph of the map

$$a \mapsto \min\left\{\frac{\lambda_k^4(a)}{\lambda_j^4(a)}; \ k = 1, \ldots 11, \ j = 0, \ldots, k - 1\right\}. \tag{3.58}$$

It turns out that it is "almost" increasing and always below 3, meaning that there exists at least one couple of indexes j and k for which $\lambda_k^4(a)/\lambda_j^4(a) \in (1, 3)$, which is the first instability interval of I_U. We numerically saw that all the ratios $\lambda_{n+1}(a)/\lambda_n(a)$ for $n \geqslant 5$ satisfy this condition (cf. Fig. 3.6), for every $a \in (0, 1)$. The right picture in Fig. 3.8 depicts the graph of $\mathbb{E}^{\ell}_{12}(a)$: we see that

for (3.32), **the maximum of** $\mathbb{E}^{\ell}_{12}(a)$ **is attained for** $a \approx 0.9$. $\tag{3.59}$

♦ We now analyze the much simpler case of (3.35). In Fig. 3.9 we display the plot of the map $a \mapsto \mathbb{E}^{\ell}_{12}(a)$ for (3.35). It turns out that

for (3.35), **the maximum of** $\mathbb{E}^{\ell}_{12}(a)$ **is attained for** $a \approx 0.5$. $\tag{3.60}$

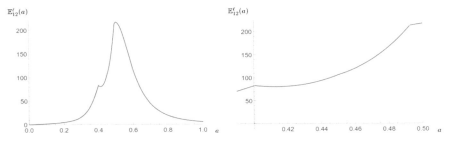

Fig. 3.9 Graph of the map $a \mapsto \mathbb{E}_{12}^{\ell}(a)$ (left), magnified for $a \in [0.4, 0.5]$ (right), for (3.35)

As expected, the minimum of the energies $E\big(\lambda(a), \rho(a)\big)$ is attained by couples of low eigenvalues, more precisely by the couple $(\lambda_0(a), \lambda_1(a))$ when $a \in (0, 0.4) \cup (0.5, 1)$ (approximately) and by the couple $(\lambda_1(a), \lambda_2(a))$ when $a \in (0.4, 0.5)$. This explains the corner close to $a = 0.4$. The reason why these are the minima is easily explained. From (3.55), we see that the minimum is necessarily reached by couples of consecutive eigenvalues (λ^4, ρ^4), possibly with a small ρ. This suggests that the minimum should always be attained by the couple (λ_0, λ_1), with some possible exception. By looking carefully at Fig. 2.1 when $a \in (0.4, 0.5)$ one understands why, in this interval, the couple (λ_1, λ_2) yields a smaller energy: the corresponding curves of eigenvalues are very close in this range.

3.4.5 Back to Nonlinear Instability: Optimal Position of the Piers

In this section, we investigate the optimal position of the piers in terms of nonlinear instability and we compare the notions of stability introduced in Definitions 3.4 and 3.6, in light of the comments written in the previous sections.

In Sect. 3.4.4, we have analyzed the linear stability for Eqs. (3.32) and (3.35). A careful look at Fig. 3.2 highlights that if we maintain fixed the ratio ρ/λ and we increase δ (that is, the energy), then we may cross some tiny instability zones with absolute value of the trace of the monodromy matrix associated with (3.44) very close to 2. Here, from a physical point of view, the instability is very weak and difficult to be detected numerically. For these reasons, we have neglected such regions in our characterization of the amplitude threshold, see (3.53); linear instability has a purely theoretical characterization, so there may be cases when its effects are not concretely visible. Even simpler is the picture in Fig. 3.3, where the tiny instability regions are empty and degenerate into double resonant lines.

For both (3.32) and (3.35), we numerically found the critical energy threshold for nonlinear instability, as characterized in Definition 3.5. The response with respect to

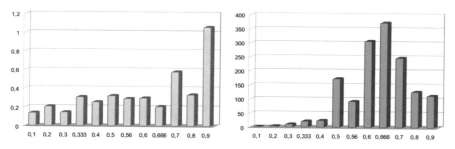

Fig. 3.10 Comparison between $a \mapsto \mathbb{E}_{12}^{\ell}(a)$ (left picture) and $a \mapsto \mathbb{E}_{12}(a)$ (right picture) for (3.32)

the optimal position of the piers has been examined following the procedure described in Sect. 4.5, in particular fixing $\eta = 0.1$. In Fig. 3.10, we deal with Eq. (3.32).

In Tables 3.1 and 3.2, we report the trace of the monodromy matrix of (3.41) and the associated expansion rates $\mathcal{ER}_{\tau}$ defined in (3.43) in correspondence of nonlinear instability, namely for the least amplitude of the prevailing mode for which (3.15) holds true (here τ is the number appearing in Definition 3.4). We recall that the trace is computed after one period of W_{λ}^{2}, that is, half a period of W_{λ}. We notice that the trace is far away from ± 2: indeed, for (3.35) the threshold of instability is obtained in correspondence of the black curve (drawn numerically) inside the region of linear instability in Fig. 3.3. An analogous observation holds for (3.32). This shows that condition (ii) in (3.15) is somehow equivalent to a "large absolute value of the trace of the monodromy matrix", namely

nonlinear instability corresponds to a sufficiently large Floquet multiplier

and is detected only if we are "sufficiently deep inside" the resonance tongues. In fact, we can make this more precise, by observing that the expansion rates reported in Table 3.2 all lie around 100. This has a simple explanation: Definition 3.4, with the choice $\eta = 0.1$, requires that, in the time interval $[0, \tau]$, the residual mode becomes 10 times larger than in the interval $[0, \tau/2]$. If the considered equation were linear, by the definition of expansion rate in (3.43) we would have $\mathcal{ER}_{\tau/2} = \sqrt{\mathcal{ER}_{\tau}}$ and in order to fulfill the second condition in (3.15) one should have

$$\sqrt{\mathcal{ER}_{\tau}} = \frac{\mathcal{ER}_{\tau}}{\mathcal{ER}_{\tau/2}} = 10,$$

so that $\mathcal{ER}_{\tau} = 100$. Of course, the presence of a nonlinear term alters the value of $\mathcal{ER}_{\tau}$, but the general principle remains true also for the nonlinear equation (3.1). Hence,

**the expansion rate (3.43) provides a fairly precise measure
of nonlinear instability.**

Table 3.1 Traces of the monodromy matrix in correspondence of nonlinear instability

a	0.1	0.2	0.3	1/3	0.4	0.5	0.56	0.6	2/3	0.7	0.8	0.9	No piers
Equation (3.32)	−2.144	−2.249	−2.397	−2.492	−2.465	−2.656	−2.465	−2.525	2.555	2.515	2.459	2.471	−2.475
Equation (3.35)	−2.047	−2.085	−2.08	−2.069	−2.068	−2.079	−2.081	−2.088	−2.101	−2.103	−2.128	−2.16	−2.289

Table 3.2 Expansion rates in correspondence of nonlinear instability

a	0.1	0.2	0.3	1/3	0.4	0.5	0.56	0.6	2/3	0.7	0.8	0.9	No piers
Equation (3.32)	123.548	75.831	65.793	96.85	88.532	179.408	90.627	111.555	119.632	102.219	81.17	86.2	617.994
Equation (3.35)	91.828	74.486	108.586	80.212	127.101	97.985	111.482	136.453	101.399	108.093	89.127	69.748	96.017

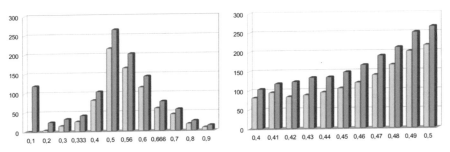

Fig. 3.11 Comparison between $a \mapsto \mathbb{E}_{12}^{\ell}(a)$ (brighter) and $a \mapsto \mathbb{E}_{12}(a)$ (darker) for problem (3.35)

We also tested other choices of η, such as $\eta = 1/15$, $\eta = 0.08$, $\eta = 0.125$, always approximately finding an expansion rate equal to $1/\eta^2$ (see Tables 3.12 and 3.13 in Sect. 3.7).

Nonlinear instability thus seems more in line with reflecting the occurrence of an abrupt and significant phenomenon inside the considered structure, so that

> **Definition 3.4 is more application-oriented and thus more suitable for practical analysis.**

In Fig. 3.11 we deal with (3.35), comparing the values of $\mathbb{E}_{12}^{\ell}(a)$ (bright) and $\mathbb{E}_{12}(a)$ (dark) for $a \in (0, 1)$ (left picture) and showing that the behaviors of $\mathbb{E}_{12}(a)$ and $\mathbb{E}_{12}^{\ell}(a)$ for $a \in (0.4, 0.5)$ are qualitatively the same (right picture). In this case,

> **the optimal values of a for (3.35) are the same**
> **with both definitions of instability.**   (3.61)

The unexpected large value of $\mathbb{E}_{12}(0.1)$ has an explanation, as well; since the corresponding ratio $\lambda_1/\lambda_0 \approx 1.185$ is close to 1, the trace of the monodromy matrix grows very slowly as δ increases, therefore it requires a large amount of energy in order to reach a sufficiently large trace yielding nonlinear instability. Basically, the black curve in Fig. 3.3 has an asymptote for $\rho/\lambda = 1$.

Summarizing, we may conclude that

> **linear instability is a clue for the occurrence of nonlinear instability,**   (3.62)

the latter requiring larger energies to occur. In particular, if there is no linear instability (as in the case $\rho < \lambda$, see both Figs. 3.2 and 3.3), one should not expect nonlinear instability.

Beyond the features illustrated so far, there is another crucial concern for choosing the definition of instability to be used. When the modes are mixed and there is no possibility for a bi-modal solution to exist, one cannot define linear instability restricting the attention on a 2×2 system. On the contrary, nonlinear instability may be checked also in cases where no nontrivial invariant subspaces exist for the considered equation. Thus, Definition 3.4 is sufficiently flexible for practical purposes and, in the sequel, we will always stick to it when speaking about instability.

3.5 Equations Which Tend to Mix the Modes

3.5.1 How Do the Modes Mix?

In this section, we discuss particular situations where the modes are mixed by the nonlinearity, thereby displaying a relevant physiological energy transfer. For simplicity, we only analyze the case of Eq. (3.1) with $\gamma_1 = \gamma_3 = 0$, leading to

$$u_{tt} + u_{xxxx} - \gamma_2 \|u_x\|_{L^2}^2 u_{xx} + f(u) = 0. \tag{3.63}$$

Let us first quickly comment on the nonlinear equation

$$u_{tt} + u_{xxxx} - \|u_x\|_{L^2}^2 u_{xx} = 0 \qquad x \in I, \quad t > 0, \tag{3.64}$$

that corresponds to (3.63) with $\gamma_2 = 1$ and $f \equiv 0$, thereby fitting in the second case of (3.13). According to Definition 3.1, a weak solution u of (3.64) satisfies $u \in C^0(\mathbb{R}_+; V(I)) \cap C^1(\mathbb{R}_+; L^2(I)) \cap C^2(\mathbb{R}_+; V'(I))$ and

$$\langle u_{tt}, v \rangle_V + \int_I u_{xx} v'' + \|u_x\|_{L^2}^2 \int_I u_x v' = 0 \qquad \forall v \in V(I), \quad t > 0.$$

Although (3.64) strongly resembles (3.32), its solutions behave quite differently since they tend to mix the modes, thus reducing the number of invariant subspaces, see Proposition 3.2. For this reason, the study of two-modes systems is useless, since two-modes solutions do not satisfy (3.64). This is also related to the discussion which will be carried out in Sect. 4.1, where it will be shown that the operator L defined on $V(I)$ by $\langle Lu, v \rangle_V = \int_I u'' v''$ is not the square of the operator $\mathcal{L}$ defined on $W(I) = \{u \in H_0^1(I); u(\pm a\pi) = 0\}$ by $\langle \mathcal{L}u, v \rangle_W = \int_I u' v'$ (notice that the pointwise constraints make sense also in $W(I)$, due to the embedding into continuous functions).

Let now

$$\Delta_{\lambda, \nu} := \int_I e_\lambda' e_\nu' \quad \text{and} \quad \Upsilon_\lambda := \Delta_{\lambda, \lambda}.$$

Due to the expression of the eigenfunctions recalled in Chap. 2, one has that

- $|\Delta_{\lambda, \nu}| \leqslant \Upsilon_\lambda \Upsilon_\nu$ by the Hölder inequality
- $\Delta_{\lambda, \nu} = 0$ if e_λ and e_ν have different parities
- $\Delta_{\lambda, \nu} = 0$ if e_λ and e_ν both belong to $C^\infty(I)$.

Here e_λ and e_ν are L^2-normalized eigenfunctions of (2.8) relative to the eigenvalues $\mu = \lambda^4$ and $\mu = \nu^4$. If we take initial data $u_0 \in V(I)$ and $u_1 \in L^2(I)$, then the solution of (3.64) can be written in Fourier series as

$$u(x, t) = \sum_\lambda \varphi_\lambda(t) e_\lambda(x),$$

where the Fourier coefficients φ_λ satisfy the infinite-dimensional system of ODEs:

$$\ddot{\varphi}_\lambda(t) + \lambda^4 \varphi_\lambda(t) + \left[\sum_\rho \Upsilon_\rho^2 \varphi_\rho(t)^2 + 2 \sum_{\nu > \rho} \Delta_{\nu,\rho} \varphi_\rho(t) \varphi_\nu(t) \right] \cdot \left[\sum_\rho \Delta_{\lambda,\rho} \varphi_\rho(t) \right] = 0$$

(3.65)

which shows that the modes are "mixed up" having strong interactions with each other. In particular,

even if some mode is initially at rest, it may become nontrivial as t varies and no solution of (3.64) with just one active mode (that is, in the form (3.27)) exists.

We then consider the case where the nonlinearity acts as a local restoring force (namely, $\gamma_1 = \gamma_2 = \gamma_3 = 0$ in (3.1)). Motivated by the nonlinearities appearing in the dynamics of suspension bridges [4, 29], a possible simplified choice is given by $f(u) = u^3$; this leads to the equation

$$u_{tt} + u_{xxxx} + u^3 = 0, \qquad x \in I, \quad t > 0, \tag{3.66}$$

complemented with the boundary and internal conditions (3.4). According to Definition 3.1, a weak solution u of (3.66) satisfies $u \in C^0(\mathbb{R}_+; V(I)) \cap C^1(\mathbb{R}_+; L^2(I)) \cap C^2(\mathbb{R}_+; V'(I))$ and the equality

$$\langle u_{tt}, v \rangle_V + \int_I \left(u_{xx} v'' + u^3 v \right) = 0 \qquad \forall v \in V(I), \quad t > 0.$$

Moreover, $u \in C^2(\bar{I} \times \mathbb{R}_+)$ and $u_{xx}(-\pi, t) = u_{xx}(\pi, t) = 0$ for all $t > 0$, see Proposition 3.1. By taking initial data

$$u(x, 0) = u_0(x) = \sum_{n \in \mathbb{N}} \alpha_n e_n, \quad u_t(x, 0) = u_1(x) = \sum_{n \in \mathbb{N}} \beta_n e_n,$$

every weak solution of (3.66) can be expanded in Fourier series as

$$u(x, t) = \sum_{n \in \mathbb{N}} \varphi_n(t) e_n(x),$$

and we consider the infinite-dimensional system of ODEs obtained by projecting (3.66) onto the eigenspace spanned by each eigenfunction e_n:

$$\ddot{\varphi}_n(t) + \lambda_n^4 \varphi_n(t) + A_n \varphi_n(t)^3 + \left(\sum_{\substack{m \in \mathbb{N} \\ m \neq n}} B_{n,m} \varphi_m(t)\right) \varphi_n(t)^2$$

$$+\left(\sum_{\substack{m,p \in \mathbb{N} \\ m,p \neq n}} C_{n,m,p} \varphi_m(t) \varphi_p(t)\right) \varphi_n(t) + D_n(t) = 0, \qquad (3.67)$$

for $n \in \mathbb{N}$. Here

$$A_n = \int_I e_n^4, \quad B_{n,m} = 3 \int_I e_n^3 e_m, \quad C_{n,m,p} = 3 \int_I e_n^2 e_m e_p, \qquad (3.68)$$

and D_n does not contain φ_n. It is worthwhile noticing that all the constants in (3.68) vary with n. This is not the case in hinged beams without piers, for which the equivalent infinite-dimensional ODE system simply reads

$$\ddot{\varphi}_n(t) + \lambda_n^4 \varphi_n(t) + \frac{3}{4} \varphi_n(t)^3 - \frac{3}{4} \varphi_{3n}(t) \varphi_n(t)^2 + C_n(t) \varphi_n(t) + \mathcal{D}_n(t) = 0,$$

see [16, Lemma 16]: these equations should be compared with (3.67). However, in view of the symmetry of the interval I, we immediately see that the only possible nonzero terms in D_n are the products of the kind $\varphi_m \varphi_p \varphi_q$, with $m + p + q + n$ even. As a consequence, the following statement holds.

Proposition 3.6 *Let* $u_0(x) = \alpha_j e_j$, $u_1(x) = \beta_j e_j$ *for some* $j \in \mathbb{N}$. *Then, there exist* C^2-*functions* φ_n, *infinitely many of which not identically zero, such that the solution of* (3.66) *is given by*

$$u(x,t) = \sum_{(n-j) \bmod 2 = 0} \varphi_n(t) e_n(x).$$

This result can be deduced from the uniqueness for the ODE problem, by noticing that, under the assumptions of Proposition 3.6, $\varphi_n \equiv 0$ is a solution of (3.67) for every n such that $(n - j) \bmod 2 \neq 0$. In particular, recalling the discussion after Definition 3.4, Proposition 3.6 states that

for (3.64) and (3.66), physiological energy transfers may occur

only between modes with the same parity.

3.5.2 Optimal Position of the Piers in Beams

In this section, we numerically evaluate the energy thresholds of nonlinear instability of (3.64) and (3.66) for different values of a, comparing them with the ones for a free

Table 3.3 Energy thresholds of instability for (3.71)

a	0.1	0.2	0.3	1/3	0.4	0.5	0.56	0.6	2/3	0.7	0.8	0.9	No piers
$\mathbb{E}_{12}(a)$	19.8	10.9	7.5	11.3	17.3	35.2	49.6	77.4	143.4	168.1	832.7	659	198.3

beam with no internal piers (see [16]). In view of the observations in the previous section, the study of linear stability for Eq. (3.63) does not make sense, because two-modes systems are not approximations of the original problem. We are thus forced to study only the nonlinear instability. A first problem in this direction is represented by the definition of the Wagner time T_W, which has been introduced in (3.40) for Eqs. (3.32) and (3.35) and deeply relies on the existence of uni-modal solutions. However, we proceed similarly by linearizing the solutions of (3.63) with j-th prevailing mode around the approximate solution which has only the j-th component, obtaining the Duffing equations

$$\ddot{W}_{\lambda_n} + \lambda_n^4 W_{\lambda_n} + \Upsilon_{\lambda_n}^4 W_{\lambda_n}^3 = 0, \quad W_{\lambda_n}(0) = \delta, \quad \dot{W}_{\lambda_n}(0) = 0 \tag{3.69}$$

for (3.64), where Υ_λ has been defined in the previous section, and

$$\ddot{W}_{\lambda_n} + \lambda_n^4 W_{\lambda_n} + A_n W_{\lambda_n}^3 = 0, \quad W_{\lambda_n}(0) = \delta, \quad \dot{W}_{\lambda_n}(0) = 0 \tag{3.70}$$

for (3.66), with A_n as in (3.68). The Wagner time $T_W = T_W(\delta)$ for (3.64) (resp., for (3.66)) is then defined as the period of the solution of (3.69) (resp., (3.70)). It can be computed via a formula similar to (3.39).

To analyze the nonlinear instability, we truncate systems (3.65) and (3.67) at the order $N = 12$, as in (3.17)–(3.18), and numerically integrate the resulting finite-dimensional systems

$$\ddot{\varphi}_n^A(t) + \lambda_n^4 \varphi_n^A(t) + \left[\sum_{\substack{m=0 \\ m \neq n}}^{11} \Upsilon_{\lambda_m}^2 \varphi_m^A(t)^2 + \sum_{\substack{m,p=0 \\ m \neq p}}^{11} \Delta_{\lambda_p, \lambda_m} \varphi_{\lambda_m}^A(t) \varphi_{\lambda_p}^A(t) \right] \cdot \left[\sum_{\substack{m=0 \\ m \neq n}}^{11} \Delta_{\lambda_n, \lambda_m} \varphi_{\lambda_m}^A(t) \right] = 0 \tag{3.71}$$

$$\ddot{\varphi}_n^A(t) + \lambda_n^4 \varphi_n^A(t) + A_n \varphi_n^A(t)^3 + \left[\sum_{\substack{m=0 \\ m \neq n}}^{11} B_{n,m} \varphi_m^A(t) \right] \varphi_n^A(t)^2 + \left[\sum_{\substack{m,p=0 \\ m,p \neq n}}^{11} C_{n,m,p} \varphi_m^A(t) \varphi_p^A(t) \right] \varphi_n^A(t) + D_n(t) = 0,$$
$$\tag{3.72}$$

for $n = 0, \ldots, 11$, where $u^A(x, t) = \sum_{n=0}^{11} \varphi_n^A(t) e_n(x)$. For every prevailing mode j, we determine the j-th energy threshold $E_j(a)$; the performed numerical analysis suggests that $E_j(a) < +\infty$ for every j and therefore that $a \mapsto \mathbb{E}_{12}(a)$ is continuous, see Theorem 3.1.

In Table 3.3, we display the energy thresholds of nonlinear instability obtained for (3.71), see Fig. 3.12 (where "no" denotes the case with no piers) and Table 3.14 in Sect. 3.7; we will further comment on them in Sect. 3.6.

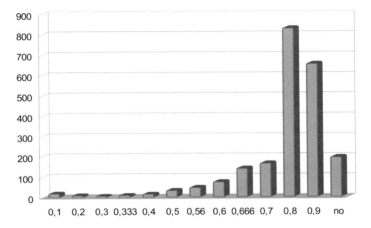

Fig. 3.12 Energy thresholds of nonlinear instability for (3.71) on varying of a

In the right picture of Fig. 3.13, we deal with system (3.72) and we draw an approximation of the graph of the function $a \mapsto \mathbb{E}_{12}(a)$, where we exploit the data reported in Table 3.17 (in Sect. 3.7). We notice that the resulting graph reaches two peaks, one around 0.35 and the other around 0.5, the latter being the optimal choice of a for the beam to remain stable. Making a comparison with the instability threshold for the beam without piers, represented by the horizontal line, we deduce that

$$\textbf{for (3.72), the piers make the beam more stable} \atop \textbf{unless the central span is much shorter than the lateral ones.} \tag{3.73}$$

Looking at Table 3.17, we also notice that $\mathbb{E}_{12}(a)$ is given either by $E_1(a)$ or by $E_2(a)$, namely the most fragile mode is either the second or the third. We can be slightly more precise in this characterization by looking at the left picture of Fig. 3.13, where we plot the two graphs (obtained via interpolation from Table 3.17) of $a \mapsto E_1(a)$ and $a \mapsto E_2(a)$. Analyzing the shape of the third eigenfunction for $a \approx 0.354$ and $a \approx 0.517$ (see Fig. 3.14), corresponding to $E_1(a) \approx E_2(a)$, we notice that

for (3.72), the second mode is the most fragile, unless the third has a zero which is "very close" to the piers.

However, this closeness has to be measured in an asymmetric way since a zero on the central span may be considered close to a pier even if its distance from it is larger. This suggests that zeros on the central span are more subject to instability than zeros on the lateral spans. The interested reader may have a look at the numerical results and at the discussion in Sect. 3.7.

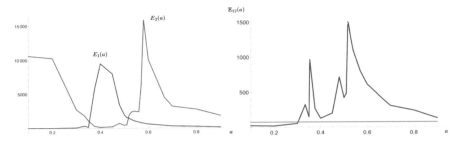

Fig. 3.13 On the left, the approximate graphs of $a \mapsto E_1(a)$ and $a \mapsto E_2(a)$. On the right, the approximate energy threshold of instability for (3.72) on varying of a, given by $\min\{E_1(a), E_2(a)\}$

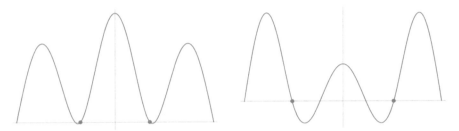

Fig. 3.14 The shape of the third eigenfunction for $a = 0.354$ (left) and $a = 0.517$ (right)

3.6 Comparing the Nonlinearities

We discuss here the information collected in the previous sections, seeking which one among the four energies

$$\mathcal{N}_1(u) = \tfrac{1}{4}\Big(\textstyle\int_I u_{xx}^2\Big)^2, \qquad \mathcal{N}_2(u) = \tfrac{1}{4}\Big(\textstyle\int_I u_x^2\Big)^2,$$
$$\mathcal{N}_3(u) = \tfrac{1}{4}\Big(\textstyle\int_I u^2\Big)^2, \qquad \mathcal{N}_4(u) = \tfrac{1}{4}\textstyle\int_I u^4,$$

introduced in Chap. 1, appears more suitable for settling a model which describes the behavior of suspension bridges. In this respect, we use three criteria to evaluate the suitability of each energy:

(a) we expect the results for the optimal position of the piers to be sufficiently robust so as not to depend on the notion of instability considered. In particular, we aim at obtaining the same optimal a using both Definitions 3.4 and 3.6 (when available), expecting claim (3.62) to be fully respected;

(b) a nonlinearity with *small physiological energy transfers* between modes (recall the discussion in Sect. 3.3) is more in line with what is observed in actual bridges: as already remarked, from [1] we know that "*one mode of oscillation prevailed*" and hence the oscillations of a suspension bridge appear in general close to a pure mode;

(c) we expect the optimal a to lie in the physical range (1.1), representing the proportion between spans which is generally respected in the construction of real bridges. Moreover, the presence of the piers should generally improve the stability of the structure.

Let us first focus on the two energies $\mathcal{N}_1$ and $\mathcal{N}_3$, studied in Sect. 3.4; we have seen in Proposition 3.2 that for both of them no physiological energy exchanges occur, since all the subspaces are invariant. However, their behavior with respect to linear stability is fairly different–see Figs. 3.2 and 3.3, where the instability tongues are depicted. Proposition 3.4 states that for $\mathcal{N}_3$ there is always linear instability if the initial amplitude δ of the prevailing mode is sufficiently large, contrary to what happens for $\mathcal{N}_1$, see Theorem 3.2.

Concerning nonlinear instability, Figs. 3.10 and 3.11 bring strong arguments in favor of (3.35) and against (3.32). In particular, Fig. 3.10 shows that for (3.32) it is $\mathbb{E}_{12}^{\ell} \ll \mathbb{E}_{12}$, namely there is a weak correlation between linear and nonlinear instability, contrary to (3.35) (apart from the case $a = 0.1$, which has been explained in Sect. 3.4.5). Taking also into account (3.61), we thus see that the above criterion a) is fulfilled only for (3.35). Furthermore, only for (3.35) the optimal values of a for linear and nonlinear instability both lie in the physical range (compare (3.59) with (3.60)). Finally, from Table 3.7 in Sect. 3.7 we infer that criterion c) is not satisfied for (3.32). Summarizing, we can conclude that

the nonlinearity $\|u\|_{L^2}^2 u$ provides a better description for the nonlocal behavior of a bridge with piers than the nonlinearity $\|u_{xx}\|_{L^2}^2 u_{xxxx}$.

In Sect. 3.5 we analyzed the energies $\mathcal{N}_2$ and $\mathcal{N}_4$, which behave in a much more complex way and, as we have seen, do not allow two-modes solutions of the considered problem. As for $\mathcal{N}_2$, the numerical responses for nonlinear instability reported in Table 3.3 are not in line with criterion c), both for the optimal value of a equal to 0.8 and for the fact that a beam without piers mostly appears more stable than with piers (except for $a = 0.8$ and $a = 0.9$). It is thus more difficult to control the stability if the stretching effects can move across the piers due to the sliding of the cable (namely, multiplying u_{xx} by the integral of u_x^2 on the whole I). Criterion c) is instead satisfied by $\mathcal{N}_4$, as we can infer by looking at Fig. 3.13 (see also (3.73)). Hence,

the nonlinearity u^3 better describes the behavior of a bridge with piers than the nonlinearity $\|u_x\|_{L^2}^2 u_{xx}$.

However, Table 3.4, together with Proposition 3.6, highlights the relevance of physiological energy transfers for $\mathcal{N}_4$. Indeed, the striking differences between the 2-modes and the 12-modes systems suggest to expect "significant physiological spread of energy" in the infinite-dimensional system. Even more, Proposition 3.6 states that

Table 3.4 Nonlinear instability for the local problem with $f(u) = u^3$: 12 vs 2 modes ($\mathbb{E}_{12}(a)$ vs $\mathbb{E}_*(a)$, see Sect. 3.7)

a	0.1	0.2	0.3	1/3	0.4	0.5	0.56	0.6	2/3	0.7	0.8	0.9	No piers
$\mathbb{E}_{12}(a)$	23.1	17.7	52.1	319.2	126.8	414.1	878.6	611.6	403.2	299.1	230.3	127.5	73.4
$\mathbb{E}_*(a)$	22.4	17.7	37.9	84.7	132.2	510	448.7	332.3	196.2	151.3	76.9	43.2	8.1

Table 3.5 Energy thresholds of nonlinear instability for problem (3.35)

a	0.1	0.2	0.3	1/3	0.4	0.5	0.56	0.6	2/3	0.7	0.8	0.9	No piers
$\mathbb{E}(a)$	120.4	25.7	34.1	42.6	104.9	263.8/265.6	203.3	144.3	78.5	57.9	26.9	14.8	3.8

the modes having the same parity as the prevailing one are physiologically affected by the nonlinearity and can significantly grow, even if not abruptly (as also observed numerically). Criterion (b) above is instead satisfied by $\mathcal{N}_3$, as shown by Table 3.5, where only one threshold is reported because we always found the same values of the critical energy except for $a = 0.5$ (where the difference, probably due to a small numerical tolerance, is anyway not significant).

Summarizing, criteria (a), (b) and (c) are only satisfied in case of equation (3.35) and we can thus conclude that

among the above nonlinearities, the nonlocal term $\|u\|_{L^2}^2 u$ gives a better description of the behavior of real structures with piers.

For this reason, when analyzing a degenerate plate model in the next section, the nonlinear restoring force due to the cables will be modeled by a suitable version of $\mathcal{N}_3$.

3.7   Tables and Numerical Results

In this section, we report with more details some of the numerical results presented in the previous sections and briefly comment on each of them. To ease their readability, we here list some notation: $\mathbb{E}_{12}^\ell$ and $\mathbb{E}_{12}$ will respectively denote the energy thresholds of linear (when defined) and nonlinear instability, as in (3.56) and (3.20). The energy threshold of nonlinear instability for a 2-modes approximation will instead be denoted by $\mathbb{E}_*$. The critical amplitude for nonlinear instability will be denoted by δ, while τ and T_W will respectively indicate the time needed to observe instability and the Wagner time, cf. Definition 3.4 and (3.40). Finally, e_j and e_k will respectively denote the eigenfunctions corresponding to the prevailing and to the residual mode fulfilling Definition 3.4, keeping the same notation therein.

★ First, we report the numerical results obtained for Eq. (3.32), where the nonlinearity is represented by a superquadratic bending term. In Table 3.6, we display the energy thresholds of linear stability, while in Tables 3.7 and 3.8 we compare the energy and the amplitude thresholds of nonlinear instability found numerically by integrating the system with 2 and with 12 modes, respectively. As described also in Sect. 4.5, the initial kinetic datum is taken identically equal to zero, while the potential one is taken equal to 0.01 on each residual mode. We complement the information about the mentioned thresholds with some more details, in particular with the value of $\delta\lambda^2$ in correspondence of nonlinear instability—in order to make a comparison with Table 3.6, cf. also Fig. 3.2. We notice that, though the couple (j, k) is not the same passing from linear to nonlinear instability, yet $\delta\lambda^2$ is always larger for nonlinear instability, confirming claim (3.62). We also observe that higher modes seem to need less time to fulfill Definition 3.4; in any case, in all our experiments it is $T_W < \tau/2$, meaning that in the determination of the thresholds we are not misled by physiological energy exchanges.

Table 3.6 Parameters of linear instability for (3.32) with 2 modes

a	0.1	0.2	0.3	1/3	0.4	0.5	0.56	0.6	2/3	0.7	0.8	0.9
$\mathbb{E}_{12}^l$	0.142	0.211	0.149	0.313	0.259	0.326	0.296	0.303	0.212	0.583	0.340	1.064
Ratio	λ_7/λ_6	λ_6/λ_5	λ_{10}/λ_9	λ_7/λ_6	λ_5/λ_4	λ_{10}/λ_9	λ_8/λ_7	λ_9/λ_8	$\lambda_{11}/\lambda_{10}$	λ_5/λ_4	λ_9/λ_8	λ_{10}/λ_9
$\delta\lambda^2$	0.503	0.599	0.514	0.708	0.654	0.72	0.692	0.699	0.6	0.909	0.733	1.137

Table 3.7 Parameters of nonlinear instability for (3.32) with 2 modes

a	0.1	0.2	0.3	1/3	0.4	0.5	0.56	0.6	2/3	0.7	0.8	0.9	No piers
$\mathbb{E}_*$	2.139	4.837	11.708	22.577	25.213	174.011	93.632	308.225	373.834	248.404	128.284	114.644	1483.48
δ	0.87	0.98	1.1	1.3	0.89	1.26	0.72	0.75	6.06	5.87	6.02	6.91	34.88
e_j	e_0	e_0	e_0	e_0	e_1	e_2	e_2	e_3	e_0	e_0	e_0	e_0	e_0
e_k	e_1	e_1	e_1	e_1	e_2	e_3	e_3	e_4	e_1	e_1	e_1	e_1	e_1
$\delta\lambda^2$	1.446	1.873	2.432	2.925	3.015	5.04	4.287	5.841	6.138	5.526	4.655	4.52	8.72
τ	15.138	7.635	4.170	3.462	2.257	1.176	0.941	0.534	3.840	4.581	6.586	8.011	15.997
T_W	2.377	1.743	1.236	1.042	0.674	0.357	0.279	0.159	1.17	1.392	1.994	2.422	3.369

♦ We then report the numerical results obtained for Eq. (3.35), where the non-linearity is given by the superquadratic L^2-term $\|u\|_{L^2}^2 u$. In Table 3.9, we display the energy thresholds of linear instability, while in Table 3.10 we report the energy thresholds of nonlinear instability found numerically. Here it is practically not necessary to distinguish between the approximations with 2 and with 12 modes, as we have already seen in Sect. 3.6. The couple (j, k) for which Definition 3.4 is fulfilled is always the same, e_0 being the prevailing mode and e_1 being the residual mode displaying instability. This is not any more true for a strictly between 0.4 and 0.5, as we have somehow already noticed with Fig. 3.9; here the couple (prevailing mode, residual mode) is given by (e_1, e_2). In Table 3.11, we thus find worth zooming on the behavior of the energy thresholds of instability for a belonging to such a range.

Once the superquadratic L^2-nonlinearity $\|u\|_{L^2}^2 u$ has been identified to be the most suitable for our analysis, we have made sure that changing the value of η appearing in Definition 3.2 does not affect the qualitative picture of our results. We chose the three values of η equal to $1/15, 0.08, 0.125$, seeking the corresponding nonlinear instability thresholds; we report the obtained values of $\mathbb{E}_*$ in Table 3.12. We chose not to display the values obtained for $a = 0.1$, since they are unrealistically large; this fact has already been given an explanation which relies on the particular position in the picture of the resonance tongues, see Fig. 3.3. In all the other cases, the qualitative behavior of the energy thresholds is the same, as well as the couple (j, k) of Definition 3.4; the time in correspondence of which instability is found ranges from 9.74 (for $a = 0.56$) to 15.995 (for $a = 0.9$), always largely more than twice the Wagner time (ranging from 1.249 for $a = 0.5$ to 2.532 for $a = 0.9$).

• In Table 3.14, we show the numerical results obtained for system (3.71), where the data do not seem to follow any clear rule and are quite far from being realistic, as we have commented in Sect. 3.6. Notice the irregular trend of the critical amplitude and the fact that, the higher the prevailing mode delivering energy is, the less is the time needed for instability to manifest.

• Finally, we dedicate some more words to (3.72). First, in Tables 3.15 and 3.16 we display the energy thresholds of instability found by analyzing the 2-modes and the 12-modes systems, respectively. We find curious that the transfer of energy is one-way (from higher to lower modes) except for $a = 0.5$ in the 2-modes system, where the first mode delivers energy to the second one.

In Table 3.17, we determine the instability thresholds for each prevailing mode up to the tenth (reporting a less precise approximation of the values), increasing δ with step 0.1. We then relate such thresholds with the position of the zeros of the corresponding eigenfunctions, observing that their distance from the piers seems indeed to play a crucial role. A careful study of the trend for odd and even modes suggests that, for a fixed, the functions $j \mapsto E_{2j}(a)$ and $j \mapsto E_{2j+1}(a)$ are increasing, up to few exceptions E_{n_i}. While looking at the shape of the corresponding eigenfunctions e_{n_i}, one sees quite clearly that their zeros not lying in the piers are considerably closer to the piers than the ones of the "preceding" eigenfunction e_{n_i-2}. We quote these two factors in Tables 3.18 and 3.19, reporting on the left the energy thresholds of Table 3.17 for even and odd modes separately, and on the right the minimal distances D_n of the zeros of e_n (not lying in a pier) from the piers. In correspondence of a

Table 3.8 Parameters of nonlinear instability for (3.32) with 12 modes

a	0.1	0.2	0.3	1/3	0.4	0.5	0.56	0.6	2/3	0.7	0.8	0.9
$\mathbb{E}_{12}$	3.087	5.699	17.378	38.131	55.172	274.462	655.22	605.881	1362.99	7175.86	274.217	216.296
δ	0.98	1.03	1.23	1.5	1.1	1.12	1.19	2.05	2.96	1.07	7.33	8.15
e_j	e_0	e_0	e_0	e_0	e_1	e_2	e_2	e_1	e_1	e_4	e_0	e_0
e_k	e_1	e_1	e_1	e_1	e_2	e_3	e_3	e_5	e_4	e_{10}	e_1	e_1
$\delta\lambda^2$	1.628	1.969	2.719	3.375	3.726	5.67	7.085	6.945	8.534	12.977	5.668	5.332
τ	15.236	8.585	4.719	3.005	1.848	0.836	0.749	2.093	2.448	6.214	7.013	8.802
T_W	2.206	1.682	1.127	0.919	0.558	0.252	0.173	0.31	0.298	0.046	1.654	2.073

Table 3.9 Parameters of linear instability for (3.35) with 2 modes

a	0.1	0.2	0.3	1/3	0.4	0.5	0.56	0.6	2/3	0.7	0.8	0.9
$\mathbb{E}_{12}^{\ell}$	1.678	4.737	15.88	27.033	82.586	216.953	166.066	115.552	60.592	44.383	19.017	9.326
ratio	1.185	1.277	1.456	1.642	2.679	6.556	7.918	8.128	8.102	8.033	7.808	7.652
δ	1.012	1.424	2.112	2.55	3.793	5.207	4.908	4.487	3.817	3.53	2.853	2.386

Table 3.10 Parameters of nonlinear instability for problem (3.35)

a	0.1	0.2	0.3	1/3	0.4	0.5	0.56	0.6	2/3	0.7	0.8	0.9	No piers
$\mathbb{E}_{12}$	120.44	25.73	34.17	42.61	104.93	263.8/265.6	203.35	144.34	78.53	57.94	26.92	14.84	3.88
δ	4.4	2.67	2.79	2.99	4.08	5.49/5.5	5.18	4.76	4.09	3.79	3.13	2.7	1.97
τ	15.997	15.652	15.989	15.364	14.421	10.6/10.55	11.345	12.36	12.665	13.701	14.304	14.082	15.964
T_W	1.534	2.109	1.93	1.844	1.552	1.277	1.368	1.492	1.737	1.874	2.27	2.636	3.721

Table 3.11 Parameters of nonlinear instability for (3.35), with a ranging from 0.4 to 0.5

a	0.4	0.41	0.42	0.43	0.44	0.45	0.46	0.47	0.48	0.49	0.5
$\mathbb{E}_{12}$	104.932	119.12	123.761	134.277	135.221	148.216	166.302	190.427	212	250.928	265.677
δ	4.08	3.59	3.6	3.67	3.64	3.73	3.86	4.03	4.17	4.42	5.5
τ	14.421	15.335	15.134	14.764	15.939	15.604	15.211	14.787	15.665	15.189	10.554
T_W	1.552	1.354	1.334	1.306	1.294	1.266	1.235	1.202	1.177	1.141	1.275

Table 3.12 The energy threshold $\mathbb{E}_*$ for the two-modes problem (3.37), for different choices of η

$\eta \downarrow a \rightarrow$	0.2	0.3	1/3	0.4	0.5	0.56	0.6	2/3	0.7	0.8	0.9
1/15	71.23	43.94	56.68	139.38	321.87	251.59	188.89	102.74	77.37	38.54	21.89
0.08	42.32	35.98	48.79	117.43	289.85	226.49	159.83	90.02	67.77	30.68	17.51
0.125	21.68	29.81	38.19	98.27	246.45	190.36	134.3	71.53	53.46	24.41	13.24

Table 3.13 The amplitude threshold of instability δ for (3.35), for different choices of η

$\eta \downarrow a \rightarrow$	0.2	0.3	1/3	0.4	0.5	0.56	0.6	2/3	0.7	0.8	0.9
1/15	3.69	3.04	3.29	4.44	5.79	5.48	5.11	4.39	4.09	3.44	2.99
0.08	3.14	2.84	3.13	4.22	5.63	5.33	4.89	4.24	3.95	3.24	2.82
0.125	2.52	2.66	2.88	4	5.39	5.09	4.67	3.99	3.71	3.05	2.62

Table 3.14 Parameters of nonlinear instability for the stretching problem (3.71)

a	0.1	0.2	0.3	1/3	0.4	0.5	0.56	0.6	2/3	0.7	0.8	0.9	No piers
$\mathbb{E}_{12}$	20.216	10.966	7.501	11.379	17.394	35.227	49.654	77.429	143.492	168.135	832.751	659.026	198.391
e_j	e_2	e_0	e_0	e_0	e_1	e_1	e_2	e_2	e_4	e_4	e_8	e_4	e_2
e_k	e_3	e_1	e_1	e_1	e_2	e_2	e_3	e_3	e_5	e_5	e_9	e_5	e_3
δ	0.97	1.69	1.3	1.47	1.33	1.46	1.27	1.41	1.21	1.17	1.11	2.31	3.4
τ	12.035	15.191	15.983	12.739	13.526	7.214	6.909	4.703	3.903	3.448	1.655	1.862	3.922
T_W	0.899	2.035	1.917	1.739	1.296	0.982	0.731	0.641	0.413	0.367	0.157	0.349	0.913

Table 3.15 Parameters of nonlinear instability for the local problem (3.66) with 2 modes

a	0.1	0.2	0.3	1/3	0.4	0.5	0.56	0.6	2/3	0.7	0.8	0.9	No piers
$\mathbb{E}_*$	22.435	17.704	37.951	84.721	132.215	510.073	448.787	332.383	196.2	151.349	76.977	43.204	8.167
δ	3.1	2.5	2.96	3.91	3.77	7.8	6.55	6.05	5.38	5.09	4.44	3.95	2.87
e_j	e_1	e_1	e_1	e_1	e_2	e_0	e_1	e_1	e_1	e_1	e_1	e_1	e_1
e_k	e_0	e_0	e_0	e_0	e_1	e_1	e_0	e_0	e_1	e_0	e_0	e_0	e_0
τ	14.52	15.97	15.17	15.05	14.75	14.76	15.9	15.53	15.59	15.91	15.3	15.74	13.83
T_W	2.721	2.534	2.048	1.783	1.418	1.307	1.274	1.361	1.569	1.689	2.062	2.446	4.024

Table 3.16 Parameters of nonlinear instability for the local problem (3.66) with 12 modes

a	0.1	0.2	0.3	1/3	0.4	0.5	0.56	0.6	2/3	0.7	0.8	0.9	No piers
$\mathbb{E}_{12}$	23.196	17.704	52.165	319.27	126.826	414.125	878.671	611.694	403.213	299.135	230.371	127.594	73.429
δ	3.14	2.5	3.38	6.38	3.7	5.16	8.33	7.5	6.9	6.44	6.41	5.67	5.58
e_j	e_1	e_1	e_1	e_1	e_2	e_2	e_1	e_1	e_1	e_1	e_1	e_1	e_1
e_k	e_0	e_0	e_0	e_0	e_1	e_1	e_0	e_0	e_0	e_0	e_0	e_0	e_0
τ	14.478	15.998	15.66	15.992	14.8	15.56	15.754	15.499	15.826	15.657	15.83	14.82	15.996
T_W	2.707	2.534	1.976	1.437	1.422	1.081	1.134	1.22	1.373	1.489	1.667	1.98	2.488

Table 3.17 Energy thresholds E_j on varying of a for (3.72)

a	E_0	E_1	E_2	E_3	E_4	E_5	E_6	E_7	E_8	E_9
0.1	619	24	10609	328	96845	3913	641741	33881	107734	676852
0.2	1352	18	10267	1938	28998	208477	13933	183748	186779	250931
0.3	3376	53	2711	33993	2206	32431	519474	29515	585070	1861480
1/3	5233	322	1856	38126	9481	28016	178051	118350	321733	1493750
0.4	2922	9481	127	14815	54148	7009	116255	478827	137040	150139
0.5	4385	1941	422	1012	36028	88890	16203	61350	182189	1447500
0.56	4770	900	2413	3560	12139	65170	154411	40426	73710	842544
0.6	5065	611	10089	1146	2070	73482	341889	270462	51749	148354
2/3	2523	403	4502	40568	3587	5866	27843	453118	418454	687344
0.7	2061	307	3224	25384	13631	6123	16503	445445	1089400	1937780
0.8	3391	240	2777	9736	45075	169085	488353	183748	89404	176802
0.9	2375	129	1850	6716	33116	82774	234085	591588	1421620	2622600
No piers	1160	74	194	1697	9741	35347	107364	300251	688532	1261680

Table 3.18 Energy thresholds E_n for (3.72) and minimal distances D_n from the piers for even modes

a	E_2	E_4	E_6	E_8	D_2	D_4	D_6	D_8
0.1	10609	96845	641741	**107734**	1.498	0.997	0.729	**0.294**
0.2	10267	28998	**13933**	186779	1.257	0.48	**0.075***	0.278*
0.3	2711	**2206**	519474	585070	0.621	**0.307***	0.576*	0.422
1/3	1856	9481	178051	321733	0.325	0.491*	0.698*	0.129
0.4	127	54148	116255	137040	0.159*	0.794*	0.289	0.319*
0.5	422	36028	**16203**	182189	0.803*	0.558	**0.426***	0.345
0.56	2413	12139	154411	**73710**	1.104*	0.158	0.683	**0.07***
0.6	10089	**2070**	341889	51749	1.257*	**0.117***	0.515	**0.379***
2/3	4502	**3587**	27843	418454	1.441*	**0.651***	0.136	0.455
0.7	3224	13631	16503	1089400	1.522*	0.843*	0.091*	0.297
0.8	2777	45075	488353	**89404**	1.76*	1.077*	0.752*	**0.476***
0.9	1850	33116	234085	1421620	2.002*	1.247*	0.897*	0.697*

Table 3.19 Energy thresholds E_n for (3.72) and minimal distances D_n from the piers for odd modes

a	E_1	E_3	E_5	E_7	E_9	D_1	D_3	D_5	D_7	D_9
0.1	24	328	3913	33881	676852	0.314*	0.314*	0.314*	0.314*	0.314*
0.2	18	1938	208477	**183748**	250931	0.628*	0.628*	0.628*	**0.628***	0.266
0.3	53	33993	**32431**	29515	1861480	0.942*	0.942*	**0.483**	**0.142***	0.395*
1/3	322	38126	**28016**	118350	1493750	1.047*	1.047*	**0.184**	0.336*	0.524*
0.4	9481	14815	**7009**	478827	**150139**	1.257*	0.718	**0.284***	0.628*	**0.064**
0.5	1941	**1012**	88890	**61350**	1447500	1.571*	D	0.785*	D	0.524*
0.56	900	3560	65170	**40426**	842544	1.759*	0.44*	0.519	**0.43***	0.211
0.6	611	1146	73482	270462	**148354**	1.885*	0.718*	0.284	0.628*	**0.064***
2/3	403	40568	**5866**	453118	687344	2.094*	1.047*	**0.184***	0.336	0.524*
0.7	307	25384	**6123**	445445	1937780	2.199*	1.138*	**0.483***	0.142	0.395
0.8	240	9736	169085	183748	**176802**	2.513*	1.341*	0.893*	0.628*	**0.266***
0.9	129	6716	82774	591588	2622600	2.827*	1.54*	1.045*	0.785*	0.625*

double zero, we set $D_n = 0$ and we use the letter D. Losses of monotonicity of the functions $j \mapsto E_{2j}(a)$, $j \mapsto E_{2j+1}(a)$ are highlighted by bold numbers. In Tables 3.18 and 3.19 we have highlighted with the symbol * the situations where the zero realizing the minimal distance from the piers lies in the central span. We exclude D_0 from this analysis, since the first eigenfunction does not present zeros out of the piers. We conclude that

the functions $j \mapsto E_{2j}$ and $j \mapsto E_{2j+1}$ have a monotone increasing trend; when this fails, the zeros of the corresponding eigenfunction have got considerably closer to the piers.

Somehow the prevailing mode undergoes more the impulsive effect of the piers if one of its zeros is closer to a pier. This appears reasonable, since, in this case, a smaller perturbation may be sufficient to "move" such zero to another span. Whether the distance from the pier has a different weight if the zero belongs to the lateral or to the central span should be further investigated.

References

1. Ammann OH, von Kármán T, Woodruff GB (1941) The failure of the Tacoma Narrows Bridge. Federal Works Agency
2. Arioli G, Gazzola F (2017) Torsional instability in suspension bridges: the Tacoma Narrows Bridge case. Commun Nonlinear Sci Numer Simul 42:342–357
3. Arioli G, Koch H (2017) Families of periodic solutions for some Hamiltonian PDEs. SIAM J Appl Dyn Syst 16:1–15
4. Augusti G, Sepe V (2001) A "deformable section" model for the dynamics of suspension bridges. Part I: model and linear response. Wind Struct 4:1–18
5. Battisti U, Berchio E, Ferrero A, Gazzola F (2017) Periodic solutions and energy transfer between modes in a nonlinear beam equation. J Math Pures Appl 108:885–917
6. Berchio E, Gazzola F (2015) A qualitative explanation of the origin of torsional instability in suspension bridges. Nonlinear Anal TMA 121:54–72
7. Berti M (2007) Nonlinear oscillations of Hamiltonian PDEs. Vol 74, Progress in nonlinear differential equations and their applications. Birkhäuser, Boston
8. Bleich F, McCullough CB, Rosecrans R, Vincent GS (1950) The mathematical theory of vibration in suspension bridges. U.S. Dept. of Commerce, Bureau of Public Roads, Washington D.C.
9. Bonheure D, Gazzola F, Moreira dos Santos E (2019) Periodic solutions and torsional instability in a nonlinear nonlocal plate equation. SIAM J Math Anal 51:3052–3091
10. Burdina VI (1953) Boundedness of solutions of a system of differential equations. Dokl Akad Nauk SSSR 92:603–606
11. Burgreen D (1951) Free vibrations of a pin-ended column with constant distance between pin ends. J Appl Mech 18:135–139
12. Cazenave T, Weissler FB (1996) Unstable simple modes of the nonlinear string. Quart Appl Math 54:287–305
13. Dickey RW (1970) Free vibrations and dynamic buckling of the extensible beam. J Math Anal Appl 29:443–454
14. Duffing G (1918) Erzwungene schwingungen bei veränderlicher eigenfrequenz. F. Vieweg und Sohn, Braunschweig
15. Ferreira V, Gazzola F, Moreira dos Santos E (2016) Instability of modes in a partially hinged rectangular plate. J Differ Equ 261:6302–6340
16. Garrione M, Gazzola F (2017) Loss of energy concentration in nonlinear evolution beam equations. J Nonlinear Sci 27:1789–1827
17. Garrione M, Gazzola F (2020) Linear theory for beams with intermediate piers. Commun Contemp Math
18. Gasparetto C, Gazzola F (2018) Resonance tongues for the Hill equation with Duffing coefficients and instabilities in a nonlinear beam equation. Commun Contemp Math 20:1750022 (22 pp)
19. Gentile G, Mastropietro V, Procesi M (2005) Periodic solutions for completely resonant nonlinear wave equations with Dirichlet boundary conditions. Commun Math Phys 256:437–490
20. Gesztesy F, Weikard R (1996) Picard potentials and Hill's equation on a torus. Acta Math 176:73–107

21. Ghisi M, Gobbino M (2001) Stability of simple modes of the Kirchhoff equation. Nonlinearity 14:1197–1220
22. Goldberg W, Hochstadt H (1979) On a Hill's equation with two gaps in its spectrum. SIAM J Math Anal 10:1069–1076
23. Holubová G, Nečesal P (2010) The Fučik spectra for multi-point boundary-value problems. Electron J Differ Equ Conf 18:33–44
24. Kuksin S (1987) Hamiltonian perturbations of infinite-dimensional linear systems with imaginary spectrum. Funktsional Anal i Prilozhen 21:22–37
25. Larsen A (2000) Aerodynamics of the Tacoma Narrows Bridge - 60 years later. Struct Eng Int 4:243–248
26. Lee C (2000) Periodic solutions of beam equations with symmetry. Nonlinear Anal TMA 42:631–650
27. Liu JQ (2002) Free vibrations for an asymmetric beam equation. Nonlinear Anal TMA 51:487–497
28. Liu JQ (2004) Free vibrations for an asymmetric beam equation, II. Nonlinear Anal TMA 56:415–432
29. Plaut RH, Davis FM (2007) Sudden lateral asymmetry and torsional oscillations of section models of suspension bridges. J Sound Vib 307:894–905
30. Rabinowitz PH (1978) Free vibrations for a semilinear wave equation. Comm Pure Appl Math 31:31–68
31. Scott R (2001) In the wake of Tacoma. Suspension bridges and the quest for aerodynamic stability. ASCE, Reston
32. Smith FC, Vincent GS (1950) Aerodynamic stability of suspension bridges: with special reference to the Tacoma Narrows Bridge, Part II: mathematical analysis, investigation conducted by the Structural Research Laboratory, University of Washington–Seattle: University of Washington Press
33. Stoker JJ (1992) Nonlinear vibrations in mechanical and electrical systems. Wiley, New York
34. Wagner H (1925) Über die entstehung des dynamischen auftriebes von tragflügeln. Zeit Angew Mathematik und Mechanik 5:17–35
35. Wayne CE (1990) Periodic and quasi-periodic solutions of nonlinear wave equations via KAM theory. Commun Math Phys 127:479–528
36. Yakubovich VA, Starzhinskii VM (1975) Linear differential equations with periodic coefficients. Wiley, New York (Russian original in Izdat. Nauka, Moscow, 1972)

Chapter 4
Nonlinear Evolution Equations for Degenerate Plates

Abstract The analysis of the stability is performed for a structure of degenerate plate-type, more suitable to describe the behavior of real bridges. Both the cases of rigid and extensible hangers are taken into account, determining again the optimal position of the piers in terms of linear and nonlinear stability, with particular emphasis on the torsional modes.

Keywords Degenerate plates · Torsional modes · Invariant subspaces · Stability · Position of the piers

We here study a system of two PDEs for a degenerate plate. The first one is a beam equation, for which the underlying linear theory has been recalled in Chap. 2; the other one is a second order PDE, for which we recall in the next section the main results about the related spectral decomposition.

4.1 Some Functional Preliminaries

It was shown in [9] that the operator L defined on $V(I)$ by $\langle Lu, v \rangle_V = \int_I u''v''$ is *not* the square of the operator $\mathcal{L}$ defined on $W(I)$ by $\langle \mathcal{L}u, v \rangle_W = \int_I u'v'$, where

$$W(I) := \{u \in H_0^1(I);\ u(\pm a\pi) = 0\} \tag{4.1}$$

and $\langle \cdot, \cdot \rangle_W$ denotes the duality pairing between $W'(I)$, the dual space of $W(I)$, and $W(I)$. Notice that, since $W(I) \subset C(I)$, the pointwise constraints still make sense. Therefore, there is no coincidence between the eigenvalues of (2.7) and the squares of the ones of

$$\int_I e'w' = \mu \int_I ew \qquad \forall w \in W(I); \tag{4.2}$$

this turns to be true only if the eigenfunction w solving (4.2) is of class C^4, in which case it is also an eigenfunction of the eigenvalue problem (2.8), associated with the eigenvalue $\lambda^4 = \mu^2$.

© The Author(s), under exclusive license to Springer Nature Switzerland AG 2019   69
M. Garrione and F. Gazzola, *Nonlinear Equations for Beams and Degenerate Plates with Piers*, PoliMI SpringerBriefs,
https://doi.org/10.1007/978-3-030-30218-4_4

It is thus useful to recall some spectral results also for (4.2), contained in [9, Theorem 13].

Theorem 4.1 *The eigenvalues $\mu = \kappa^2$ of problem* (4.2) *are completely determined by the numbers $\kappa > 0$ such that*

$$(i) \quad \sin(\kappa a\pi)\sin[\kappa(1-a)\pi] = 0 \quad or \quad (ii) \quad \cos(\kappa a\pi)\sin[\kappa(1-a)\pi] = 0,$$

that is,

$$(i) \quad \kappa \in \frac{\mathbb{N}}{a} \cup \frac{\mathbb{N}}{1-a} \quad or \quad (ii) \quad \kappa \in \frac{2\mathbb{N}+1}{2a} \cup \frac{\mathbb{N}}{1-a}. \tag{4.3}$$

• *If $\kappa \notin \mathbb{N}/(1-a)$, denoting by χ_0 the characteristic function of I_0, then:*
– *in case* (i), *$\mu = \kappa^2$ is a simple eigenvalue associated with the odd eigenfunction $\mathbf{D}_\kappa(x) = \chi_0(x)\sin(\kappa x)$;*
– *in case* (ii), *$\mu = \kappa^2$ is a simple eigenvalue associated with the even eigenfunction $\mathbf{P}_\kappa(x) = \chi_0(x)\cos(\kappa x)$.*
• *If $\kappa \in \mathbb{N}/(1-a)$, then the following situations may occur:*
– *if $\kappa \notin \mathbb{N}/a$ and $\kappa \notin (2\mathbb{N}+1)/2a$, then $\mu = \kappa^2$ is a double eigenvalue associated with the eigenfunctions*

$$\mathcal{D}_\kappa(x) = \begin{cases} \sin[\kappa(x+\pi)] & \text{if } x \in \overline{I}_- \\ 0 & \text{if } x \in \overline{I}_0 \\ \sin[\kappa(x-\pi)] & \text{if } x \in \overline{I}_+, \end{cases} \qquad \mathcal{P}_\kappa(x) = \begin{cases} \sin[\kappa(x+\pi)] & \text{if } x \in \overline{I}_- \\ 0 & \text{if } x \in \overline{I}_0 \\ \sin[\kappa(\pi-x)] & \text{if } x \in \overline{I}_+, \end{cases}$$

respectively odd and even;
– *if $\kappa \in \mathbb{N}/a$, then $\mu = \kappa^2$ is a triple eigenvalue associated with $\mathcal{D}_\kappa$, $\mathcal{P}_\kappa$ and $\mathbf{D}_\kappa$;*
– *if $\kappa \in (2\mathbb{N}+1)/2a$, then $\mu = \kappa^2$ is a triple eigenvalue associated with $\mathcal{D}_\kappa$, $\mathcal{P}_\kappa$ and $\mathbf{P}_\kappa$.*

Notice that (4.3)-(i) corresponds to odd eigenfunctions, while (4.3)-(ii) to even ones. Since there are no smooth junction constraints, the eigenfunctions of (4.2) are obtained by juxtaposing the eigenfunctions belonging to H_0^1 of each span and for this reason they are not C^1, in general. It is also worthwhile mentioning that the eigenvalues may be both simple or multiple, the simple ones being always associated with eigenfunctions being zero on the side spans. In case of multiple eigenvalues, the definition of the related eigenfunctions is quite arbitrary; the above choice has been motivated by the possibility of analyzing separately the behavior on the central span, that is the most vulnerable part in bridges. In Fig. 4.1 we depict the curves of eigenvalues in the plane (a, κ): the bold hyperbolas correspond to $\kappa \in \mathbb{N}/(1-a)$, the dashed lines to $\kappa \in (2\mathbb{N}+1)/2a$ and the dot-dashed ones to $\kappa \in \mathbb{N}/a$.

Each bold hyperbola in Fig. 4.1 carries a double eigenvalue, with one even and one odd eigenfunction, while each of the dashed and dot-dashed hyperbolas therein carries a simple eigenvalue, alternating even and odd eigenfunctions. In Figs. 4.2 and 4.3 we depict the shape of the first ten/twelve eigenfunctions for $a = 1/2$ and

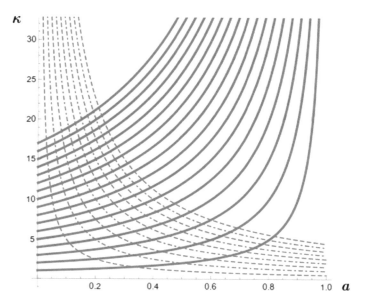

Fig. 4.1   A pictorial description of the curves of eigenvalues for (4.2), in the (a, κ)-plane

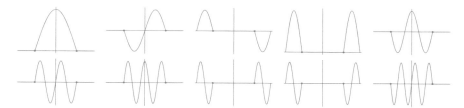

Fig. 4.2   The shape of the first ten eigenfunctions of (4.2) when $a = 14/25$

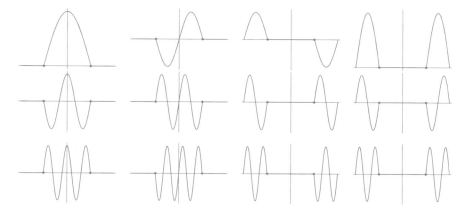

Fig. 4.3   The shape of the first twelve eigenfunctions of (4.2) when $a = 1/2$

$a = 14/25$; in case of multiple eigenvalues we plot first $\mathbf{D}_\kappa$ or $\mathbf{P}_\kappa$ (if they exist), then $\mathcal{D}_\kappa$, finally $\mathcal{P}_\kappa$.

We close this brief summary about the second order eigenvalue problem with an important observation. It was shown in [9] that the first eigenvalue of (4.2) is simple for $a > 1/3$, triple for $a = 1/3$ and double for $a < 1/3$, while the second eigenvalue is simple for $a > 1/2$ and triple for $a = 1/2$. More in general, it was observed therein that the multiplicity increases on low eigenvalues when a is small. From the point of view of nonlinear problems, since multiplicity plays against stability, we infer that

dealing with nonlinear problems it appears more convenient to take large a.

This is in line with the "physical range" (1.1).

4.2   Weak Formulation: Well-Posedness and Torsional Stability

We turn here our attention to possible suspension bridge models. A beam only possesses one degree of freedom (the vertical displacement) but, for the analysis of a bridge, it is crucial to consider a model allowing to view a torsion of the deck. In this section we propose a degenerate plate-type one, focusing on its well-posedness and on some of its qualitative properties.

In Chap. 1, we introduced system (1.8), which in adimensional form may be written as

$$\begin{cases} u_{tt} + u_{xxxx} + \left(\int_I (u^2 + \theta^2) \right) u + 2 \left(\int_I u\theta \right) \theta + f(u + \theta) + f(u - \theta) = 0 \\ \theta_{tt} - \theta_{xx} + 2 \left(\int_I u\theta \right) u + \left(\int_I (u^2 + \theta^2) \right) \theta + f(u + \theta) - f(u - \theta) = 0. \end{cases}$$

(4.4)

We complement (4.4) with the boundary-internal-initial conditions (1.9)–(1.10)–(1.11), that we rewrite here for the reader's convenience:

$$u(\pm\pi, t) = u(\pm a\pi, t) = \theta(\pm\pi, t) = \theta(\pm a\pi, t) = 0 \quad t \geqslant 0, \tag{4.5}$$

$$u(x, 0) = u_0(x), \quad u_t(x, 0) = u_1(x), \quad \theta(x, 0) = \theta_0(x), \quad \theta_t(x, 0) = \theta_1(x) \quad x \in I. \tag{4.6}$$

Similarly as in Definition 3.1, we define weak solutions as follows.

Definition 4.1 We say that the functions

$$u \in C^0(\mathbb{R}_+; V(I)) \cap C^1(\mathbb{R}_+; L^2(I)) \cap C^2(\mathbb{R}_+; V'(I))$$
$$\theta \in C^0(\mathbb{R}_+; W(I)) \cap C^1(\mathbb{R}_+; L^2(I)) \cap C^2(\mathbb{R}_+; W'(I))$$

are weak solutions of (4.4)–(4.5)–(4.6) if they satisfy the initial conditions (4.6) with $u_0 \in V(I), \theta_0 \in W(I), u_1 \in L^2(I), \theta_1 \in L^2(I)$ and if

$$\langle u_{tt}, \varphi \rangle_V + \int_I u_{xx} \varphi'' + \int_I (u^2 + \theta^2) \cdot \int_I u\varphi + 2 \int_I u\theta \cdot \int_I \theta\varphi + \int_I (f(u - \theta) + f(u + \theta))\varphi = 0,$$

$$\langle \theta_{tt}, \psi \rangle_W + \int_I \theta_x \psi' + 2 \int_I u\theta \cdot \int_I u\psi + \int_I (u^2 + \theta^2) \cdot \int_I \theta\psi + \int_I (f(u + \theta) - f(u - \theta))\psi = 0,$$

for all $(\varphi, \psi) \in V(I) \times W(I)$ and all $t > 0$ (where the spaces $V(I)$ and $W(I)$ are defined in (2.1) and (4.1), respectively).

The corresponding well-posedness result reads as follows.

Proposition 4.1 *Let $u_0 \in V(I)$, $\theta_0 \in W(I)$, $u_1, \theta_1 \in L^2(I)$. Assume that f satisfies (3.3) and $|f(s)| \leqslant C(1 + |s|^p)$ for every $s \in \mathbb{R}$ and for some $p \geqslant 1$. Then there exists a unique weak solution (u, θ) of (4.4)–(4.5)–(4.6). Moreover, $u \in C^2(\overline{I} \times \mathbb{R}_+)$ and $u_{xx}(-\pi, t) = u_{xx}(\pi, t) = 0$ for all $t > 0$.*

We omit the proof of Proposition 4.1 because it can be obtained by combining the classical proof for simple beams in [7], the proof for a full plate in [8, Theorem 3], and the proof of [11, Theorems 8 and 11], see also [4]. For the C^2-regularity of u we invoke again [12, Lemma 2.2].

Clearly, there exist solutions which have both nontrivial longitudinal and torsional components. However, if $\theta_0(x) = \theta_1(x) = 0$ (resp., $u_0(x) = u_1(x) = 0$) in (4.6), then the solution of (4.4) satisfies $\theta(x, t) \equiv 0$ (resp., $u(x, t) \equiv 0$). Hence, the above zero-initial conditions for θ or u give rise to purely longitudinal (resp., purely torsional) solutions. Since the phase space is now $V(I) \times W(I)$, according to Definition 3.3 this means that

$$\text{the spaces } V(I) \times \{0\} \text{ and } \{0\} \times W(I) \text{ are invariant for (4.4).} \qquad (4.7)$$

Our purpose is to consider solutions which are both longitudinal and torsional but that are "initially close to a purely longitudinal configuration". Therefore, we consider solutions (u, θ) of (4.4) with a *longitudinal prevailing mode*, see Definition 3.2, suitably adapted to take into account also the presence of the torsional components. In turn, this enables us to study the *stability of purely longitudinal modes*, according to the following definition.

Definition 4.2 Let $T_W > 0$. We say that a weak solution (u, θ) of (4.4) having η-prevailing longitudinal mode j, is *torsionally unstable* before time $T > 2T_W$ if there exist a torsional mode k and a time instant τ with $2T_W < \tau < T$ such that

$$\frac{\|\psi_k\|_{L^\infty(0,\tau)}}{\|\psi_k\|_{L^\infty(0,\tau/2)}} > \frac{1}{\eta}, \qquad (4.8)$$

where η is the number appearing in Definition 3.2 and ψ_k is the k-th torsional Fourier component. We say that (u, θ) is *torsionally stable* until time T if, for any torsional mode k, (4.8) is not fulfilled for any $\tau \in (2T_W, T)$.

Compared with (3.15), notice that here the indexes j and k do not refer to the same kind of modes, since the former is always longitudinal, while the latter is torsional. Furthermore, there is one condition missing in (4.8); it is not required here that the torsional mode becomes significantly large when compared to the prevailing longitudinal mode, since these two amplitudes measure two different kinds of oscillations. What really measures the torsional instability is the sudden growth of the torsional component, when compared to itself. Indeed, small torsional oscillations were never observed in suspension bridges, which means that they suddenly switch from tiny invisible oscillations to large and dangerous oscillations. This is the phenomenon under study, which we call torsional instability. In order to study the torsional stability we take advantage of some of the material developed in Chaps. 2 and 3.

4.3   The Case of Rigid Hangers

4.3.1   Linear Torsional Instability for Two-Modes Systems

As already mentioned in Chap. 1, the main contribution to the instability of suspension bridges comes from the sustaining cables. A possible simplification consists then in considering inextensible hangers, see [16]. This means that the hangers rigidly connect the deck with the cables and that the only restoring force acting on the degenerate plate is due to the cables. In this case, (4.4) becomes

$$\begin{cases} u_{tt} + u_{xxxx} + \left(\int_I (u^2 + \theta^2) \right) u + 2 \left(\int_I u\theta \right) \theta = 0 \\ \theta_{tt} - \theta_{xx} + 2 \left(\int_I u\theta \right) u + \left(\int_I (u^2 + \theta^2) \right) \theta = 0, \end{cases} \tag{4.9}$$

which has several common points with (3.35) and, therefore, enables us to exploit part of the results obtained in Sect. 3.4. From Proposition 3.2 we learn that (4.9) has many invariant subspaces and a full analysis of all the cases would be quite lengthy. Both (4.7) and Table 3.10 in Sect. 3.7 (together with the discussion therein) suggest that the case of a couple of modes, one longitudinal and one torsional, is the most meaningful and is quite representative of the whole instability phenomenon leading to an energy transfer from longitudinal to torsional modes. Therefore, we reduce again to the case of a two modes system but, contrary to (3.30), we obtain a system that *does not* necessarily solve (4.9). Let us explain this fact with full precision.

Let e_λ be an L^2-normalized eigenfunction of (2.8) related to the eigenvalue $\mu = \lambda^4$ and let η_κ be an L^2-normalized eigenfunction of (4.2) related to the eigenvalue $\mu = \kappa^2$, see Theorem 4.1. The coupling between these modes is measured by the coefficient

$$A_{\lambda,\kappa} = A_{\lambda,\kappa}(a) := \int_I e_\lambda \eta_\kappa.$$

Note that

$$A_{\lambda,\kappa}^2 < 1 \qquad (4.10)$$

in view of the Hölder inequality (recall that $e_\lambda \not\equiv \eta_\kappa$, since the latter is identically zero on some span). Moreover,

$$\text{if } e_\lambda \text{ and } \eta_\kappa \text{ have opposite parities, then } A_{\lambda,\kappa} = 0. \qquad (4.11)$$

But there are also cases where $A_{\lambda,\kappa} \neq 0$, see Table 4.1. We here choose to consider only the first two torsional modes. Thus, we stick to values of a for which such modes are simple, as underlined at the end of Sect. 4.1; the only exception is represented by $a = 0.5$, which is included since it belongs to the range (1.1). For $a = 0.5$, Theorem 4.1 states that the second eigenvalue has multiplicity 3; the corresponding value of A_{λ_n,κ_1} in Table 4.1 corresponds to the eigenfunction $\mathbf{D}_{\kappa_1}(x) = \chi_0(x)\sin(2x)$, see the second picture in Fig. 4.3.

It appears that only for the couples $(0,0)$ and $(1,1)$ the value of $A_{\lambda,\kappa}$ is relevant, especially if we take into account that the "coupling between modes" is measured by $2A_{\lambda,\kappa}^2$, as we now explain.

If $A_{\lambda,\kappa} = 0$, then the space $\langle e_\lambda \rangle \times \langle \eta_\kappa \rangle$ is invariant and one can seek solutions of (4.9) in the form

$$\Big(u(x,t), \theta(x,t) \Big) = \Big(w(t)e_\lambda(x), z(t)\eta_\kappa(x) \Big), \qquad (4.12)$$

to be compared with (3.29). By plugging (4.12) into (4.9) we see that the couple (w, z) satisfies the system

$$\begin{cases} \ddot{w}(t) + \lambda^4 w(t) + (1 + 2A_{\lambda,\kappa}^2)z(t)^2 w(t) + w(t)^3 = 0 \\ \ddot{z}(t) + \kappa^2 z(t) + (1 + 2A_{\lambda,\kappa}^2)w(t)^2 z(t) + z(t)^3 = 0, \end{cases} \qquad (4.13)$$

where, for later use, we left $A_{\lambda,\kappa}$ explicitly written even if it is zero.

If $A_{\lambda,\kappa} \neq 0$, then the space $\langle e_\lambda \rangle \times \langle \eta_\kappa \rangle$ is not invariant for (4.9). Therefore, there are no solutions of (4.9) in the form (4.12). Nevertheless, since a couple of (longitudinal, torsional) modes is quite representative of the dynamics in view of (4.7), since $A_{\lambda,\kappa}$ is quite small in most cases (see Table 4.1), and since Table 3.10 in Sect. 3.7 (together with the discussion therein) shows that the energy thresholds do not vary significantly while comparing two-modes systems with twelve-modes systems, we still focus our attention on the two-modes system (4.13). Clearly, the resulting solutions should be seen as nonexact (approximated) solutions of (4.9).

We associate with (4.13) the initial conditions

$$w(0) = \delta > 0, \quad z(0) = z_0, \quad \dot{w}(0) = \dot{z}(0) = 0 \qquad (4.14)$$

and we notice that if $z_0 = 0$, then the solution of (4.13)–(4.14) is $(w, z) = (W_\lambda, 0)$ where W_λ solves (3.36), see (3.38). Hence, W_λ has minimal period as in (3.39), so that W_λ^2 has minimal period equal to

Table 4.1 Some values of $A_{\lambda_n,\kappa_m} = A_{\lambda_n,\kappa_m}(a)$

$a \downarrow (n,m) \rightarrow$	(0,0)	(2,0)	(4,0)	(6,0)	(8,0)	(10,0)	(1,1)	(3,1)	(5,1)	(7,1)	(9,1)	(11,1)
0.5	0.953	0.034	0.007	0.002	0.001	$5 \cdot 10^{-4}$	0.5	0.476	0	0.017	0	0.003
14/25	0.979	0.005	0.012	$2 \cdot 10^{-5}$	0.001	10^{-4}	0.857	0.121	0.01	0.004	0.003	$2 \cdot 10^{-4}$
0.6	0.986	0	0.011	$8 \cdot 10^{-4}$	$6 \cdot 10^{-4}$	$4 \cdot 10^{-4}$	0.936	0.041	0.016	0	0.004	$5 \cdot 10^{-4}$
2/3	0.989	0.004	0.002	0.002	10^{-4}	$2 \cdot 10^{-4}$	0.975	0	0.019	0.002	0	0.001
0.7	0.989	0.006	$2 \cdot 10^{-4}$	0.002	$3 \cdot 10^{-4}$	10^{-5}	0.981	0.003	0.008	0.004	$5 \cdot 10^{-4}$	$2 \cdot 10^{-4}$
0.8	0.986	0.011	10^{-4}	$9 \cdot 10^{-5}$	10^{-4}	$3 \cdot 10^{-4}$	0.981	0.014	0.001	0	0.001	$7 \cdot 10^{-4}$
0.9	0.98	0.016	0.002	$4 \cdot 10^{-4}$	10^{-4}	$5 \cdot 10^{-5}$	0.970	0.023	0.003	0.001	$3 \cdot 10^{-4}$	10^{-4}

$$T_\lambda(\delta) := \frac{T(\delta)}{2} = 2\sqrt{2} \int_0^{\pi/2} \frac{d\phi}{\sqrt{2\lambda^4 + \delta^2(1 + \sin^2 \phi)}}; \qquad (4.15)$$

the function $\delta \mapsto T_\lambda(\delta)$ is continuous, decreasing and such that $T_\lambda(0) = \pi/\lambda^2$.

We then consider initial conditions (4.14) with $0 < |z_0| < \eta^2\delta$, with η small, which justifies (1.6) and the stability analysis of (4.9). For this study, we slightly modify Definition 3.6.

Definition 4.3 The λ-longitudinal mode W_λ is said to be *linearly stable (unstable)* with respect to the κ-torsional-mode if $\xi \equiv 0$ is a stable (unstable) solution of the linear Hill equation

$$\ddot{\xi}(t) + \left(\kappa^2 + (1 + 2A_{\lambda,\kappa}^2)W_\lambda(t)^2\right)\xi(t) = 0. \qquad (4.16)$$

Several remarks are in order. With this definition, we see that the linear stability *does not* depend on the positive coefficient γ in (1.8) multiplying the nonlocal terms in (4.9): up to a suitable scaling (4.16) remains the same. In fact, (4.16) *is not* a standard Hill equation with a periodic coefficient depending on two parameters measuring its mean value and its amplitude of oscillation: the presence of the coefficient $A_{\lambda,\kappa}$ prevents the use of classical methods for the study of the stability. This coefficient depends in a very hidden way on the two parameters λ and κ and is well-defined only if λ^4 is an eigenvalue of (2.8) and κ^2 is an eigenvalue of (4.2). Summarizing, there is no way, not even numerically, to determine the resonant tongues because ...they are not defined for all λ and κ! Let us recall that, from a physical point of view, $A_{\lambda,\kappa}$ depends on the coupling between the torsional and the longitudinal modes involved. Finally, since it may happen both that $A_{\lambda,\kappa} = 0$ and $A_{\lambda,\kappa} \neq 0$, for the description of the linear stability we need to distinguish two cases. The case $A_{\lambda,\kappa} = 0$ can be directly inferred from Proposition 3.4 simply by replacing ρ^2 with κ.

Proposition 4.2 *Let λ^4 and κ^2 be, respectively, eigenvalues of (2.8) and (4.2) such that $A_{\lambda,\kappa} = 0$. The λ-longitudinal-mode of (4.9) of amplitude δ is linearly stable with respect to the κ-torsional-mode if and only if one of the following facts holds:*

$$\lambda^2 > \kappa \text{ and } \delta > 0 \quad \text{or} \quad \lambda^2 < \kappa \text{ and } \delta \leqslant \sqrt{2(\kappa^2 - \lambda^4)}.$$

Since the linear instability is a clue for the nonlinear instability, Proposition 4.2 states that if $\lambda^2 > \kappa$ then (4.9) does not display torsional instability according to Definition 4.2 (in fact, for any T_W and T). The stability picture describing the situation of Proposition 4.2 may be derived from Fig. 3.3, again by replacing ρ^2 with κ.

When $A_{\lambda,\kappa} \neq 0$, it appears out of reach to obtain a statement as precise as Proposition 4.2. We are able to prove the following.

Theorem 4.2 *Let λ^4 and κ^2 be, respectively, eigenvalues of (2.8) and (4.2) such that $A_{\lambda,\kappa} \neq 0$.*

• The λ-longitudinal-mode of (4.13) of amplitude δ is linearly unstable with respect to the κ-torsional-mode whenever δ is sufficiently large.

• If $\kappa^2 \neq \lambda^4$, then the λ-longitudinal-mode of (4.13) of amplitude δ is linearly stable with respect to the κ-torsional-mode whenever δ is sufficiently small. In particular, linear stability is guaranteed whenever there exists an integer m such that both the two following conditions hold:

$$\frac{1}{2} \log \left(1 + \tfrac{3\delta^2}{\kappa^2} \right) + 2\sqrt{2} \int_0^{\pi/2} \sqrt{\frac{1 + 3(\delta^2/\kappa^2)\sin^2 \phi}{2(\lambda^4/\kappa^2) + (\delta^2/\kappa^2)(1 + \sin^2 \phi)}}\, d\phi \leq (m+1)\pi, \qquad (4.17)$$

$$2\sqrt{2} \int_0^{\pi/2} \sqrt{\frac{1 + (\delta^2/\kappa^2)\sin^2 \phi}{2(\lambda^4/\kappa^2) + (\delta^2/\kappa^2)(1 + \sin^2 \phi)}}\, d\phi - \frac{1}{2} \log \left(1 + \tfrac{3\delta^2}{\kappa^2} \right) \geq m\pi. \qquad (4.18)$$

Proof The linear instability for large δ is a consequence of the following result, essentially due to Cazenave–Weissler [6].

Lemma 4.1 *Assume that $A_{\lambda,\kappa} \neq 0$. Then the λ-longitudinal-mode of (4.9) of amplitude δ is linearly unstable with respect to the κ-torsional-mode provided that δ is sufficiently large.* $\qquad\square$

Proof For all $\varepsilon > 0$ we define Z_ε and ξ_ε by

$$W_\lambda(t) = \frac{\lambda^2}{\sqrt{\varepsilon}} Z_\varepsilon \left(\frac{\lambda^2}{\sqrt{\varepsilon}} t \right), \qquad \xi(t) = \xi_\varepsilon \left(\frac{\lambda^2}{\sqrt{\varepsilon}} t \right).$$

If W_λ satisfies (3.36), then Z_ε solves the equation

$$\ddot{Z}_\varepsilon(t) + \varepsilon Z_\varepsilon(t) + Z_\varepsilon(t)^3 = 0 \qquad (4.19)$$

and, according to Definition 4.3, the linear stability of (4.13) depends on the stability of the following Hill equation:

$$\ddot{\xi}_\varepsilon(t) + (1 + 2A_{\lambda,\kappa}^2) \left(\frac{\kappa^2 \varepsilon}{(1 + 2A_{\lambda,\kappa}^2)\lambda^4} + Z_\varepsilon(t)^2 \right) \xi_\varepsilon(t) = 0. \qquad (4.20)$$

By letting $\varepsilon \to 0$, we see that (4.19) and (4.20) "converge" respectively to the limit problems

$$\ddot{Z}(t) + Z(t)^3 = 0, \qquad \ddot{\xi}(t) + (1 + 2A_{\lambda,\kappa}^2) Z(t)^2 \, \xi(t) = 0. \qquad (4.21)$$

These limit equations are precisely (3.1) and (3.2) in [6] with $\gamma = 1 + 2A_{\lambda,\kappa}^2 \in (1,3)$ in view of (4.10). Therefore, [6, Theorem 3.1] applies and the statement follows. $\square$

Then we turn to the delicate part of Theorem 4.2, namely the stability result for small δ. We apply Lemma 3.1 to the Hill equation (4.16): we take $p(t) = \kappa^2 + (1 + 2A_{\lambda,\kappa}^2)W_\lambda(t)^2$ so that

$$\log \frac{\max p}{\min p} = \log \left(1 + \frac{(1 + 2A_{\lambda,\kappa}^2)\delta^2}{\kappa^2} \right).$$

From (3.38)–(3.39) (with $\gamma_1 = 0$ and $\gamma_3 = 1$) we know that

$$W_\lambda(t) = \delta \, \mathrm{cn} \left[t \, \sqrt{\lambda^4 + \delta^2}, \frac{\delta}{\sqrt{2(\lambda^4 + \delta^2)}} \right]$$

and that the period of W_λ^2 is given by (4.15). Moreover, from the energy conservation in (3.36) we see that

$$2\dot{W}_\lambda(t)^2 = \left(\delta^2 - W_\lambda(t)^2 \right) \left(2\lambda^4 + \delta^2 + W_\lambda(t)^2 \right);$$

therefore, thanks to symmetries and with the change of variables $W_\lambda(t) = \delta \sin \phi$, we obtain

$$\int_0^{T_\lambda(\delta)} \sqrt{\kappa^2 + (1 + 2A_{\lambda,\kappa}^2) W_\lambda(t)^2} \, dt = 2 \int_0^{T_\lambda(\delta)/2} \sqrt{\kappa^2 + (1 + 2A_{\lambda,\kappa}^2) W_\lambda(t)^2} \, dt$$

$$= 2\sqrt{2} \int_0^{\pi/2} \sqrt{\frac{\kappa^2 + (1 + 2A_{\lambda,\kappa}^2)\delta^2 \sin^2 \phi}{2\lambda^4 + \delta^2 + \delta^2 \sin^2 \phi}} \, d\phi.$$

Lemma 3.1 then states that the trivial solution of (4.16) is stable provided that there exists $m \in \mathbb{N}$ such that

$$m\pi + \tfrac{1}{2} \log \left(1 + \frac{(1+2A_{\lambda,\kappa}^2)\delta^2}{\kappa^2} \right) < 2\sqrt{2} \int_0^{\pi/2} \sqrt{\frac{\kappa^2 + (1+2A_{\lambda,\kappa}^2)\delta^2 \sin^2 \phi}{2\lambda^4 + \delta^2 + \delta^2 \sin^2 \phi}} \, d\phi$$

$$< (m+1)\pi - \tfrac{1}{2} \log \left(1 + \frac{(1+2A_{\lambda,\kappa}^2)\delta^2}{\kappa^2} \right). \tag{4.22}$$

Condition (4.22) is by far less clear than the corresponding condition in Theorem 3.2. As $\delta \to 0$ it becomes

$$m\pi + (1 + 2A_{\lambda,\kappa}^2)\frac{\delta^2}{2\kappa^2} < \pi \frac{\kappa}{\lambda^2} + \frac{\delta^2}{2\kappa^2} \left(\frac{\pi(1+2A_{\lambda,\kappa}^2)}{2} \frac{\kappa}{\lambda^2} - \frac{3\pi}{4} \left(\frac{\kappa}{\lambda^2} \right)^3 \right) < (m+1)\pi - (1 + 2A_{\lambda,\kappa}^2)\frac{\delta^2}{2\kappa^2}.$$

This is a sufficient condition for linear stability for small δ and enables us to obtain the following necessary condition for linear instability for small δ: there exists $m \in \mathbb{N}$ such that

$$\frac{\delta^2}{2\kappa^2} F_- \left(\frac{\kappa}{\lambda^2} \right) < \pi \left(\frac{\kappa}{\lambda^2} - m \right) < \frac{\delta^2}{2\kappa^2} F_+ \left(\frac{\kappa}{\lambda^2} \right),$$

where

$$F_{\pm}(s) := \frac{3\pi}{4} s^3 - \frac{\pi(1 + 2A_{\lambda,\kappa}^2)}{2} s \pm (1 + 2A_{\lambda,\kappa}^2) \qquad (s \geqslant 0).$$

Simple calculus arguments show that:

    – there exists a unique $\bar{s} > 0$ such that $F_-(s) < 0$ if $s \in [0, \bar{s})$, $F_-(\bar{s}) = 0$, $F_-(s) > 0$ if $s \in (\bar{s}, \infty)$; we notice that from (4.10) and since $A_{\lambda,\kappa} \neq 0$, we have

$$\frac{3\pi}{4} s^3 - \frac{3\pi}{2} s - 3 =: G^1(s) < F_-(s) < G^0(s) := \frac{3\pi}{4} s^3 - \frac{\pi}{2} s - 1 \qquad \forall s > 0.$$

Both G^0 and G^1 admit a unique zero. Since $G^1(2) = 3\pi - 3 > 0$ and $G^0(1) = -1 + \pi/4 < 0$, we infer that $\bar{s} \in (1, 2)$ for any value of $A_{\lambda,\kappa}$.

    – since $A_{\lambda,\kappa}^2 < 1 < \frac{1}{2}(\frac{81}{2\pi^2} - 1)$, we have that $F_+(s) > 0$ for all $s \geqslant 0$.

Therefore, the resonant tongue U_m emanating from the point $(\frac{\delta^2}{2\kappa^2}, \frac{\kappa}{\lambda^2}) = (0, m)$ necessarily lies (as $\delta \to 0$) in one of the gray regions depicted in Fig. 4.4, which occur if $\kappa/\lambda^2 \in \mathbb{N}$ and, respectively,

$$\frac{\kappa}{\lambda^2} > \bar{s} \quad \text{or} \quad \frac{\kappa}{\lambda^2} < \bar{s}. \tag{4.23}$$

Since $\bar{s} \in (1, 2)$, the only case where the second situation in (4.23) may occur is when $\kappa = \lambda^2$, which is excluded by the assumptions. Thus, only the case of the left picture in Fig. 4.4 is possible, proving the second statement of Theorem 4.2. We have so proved that the λ-longitudinal-mode of (4.9) of amplitude δ is linearly stable with respect to the κ-torsional-mode whenever δ is sufficiently small. We now make quantitative this smallness requirement: we get rid of the "mysterious term" $A_{\lambda,\kappa}$ and we obtain uniform bounds for δ not depending on it. We first notice that, since $1 + 2A_{\lambda,\kappa}^2 < 3$ in view of (4.10), the right inequality in (4.22) is certainly satisfied

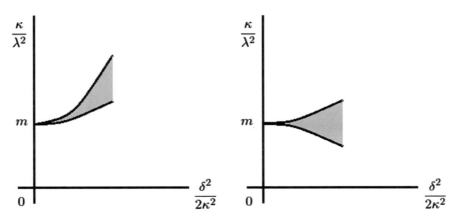

Fig. 4.4 Local bounds for the resonance tongue U_m (gray) emanating from $(0, m)$

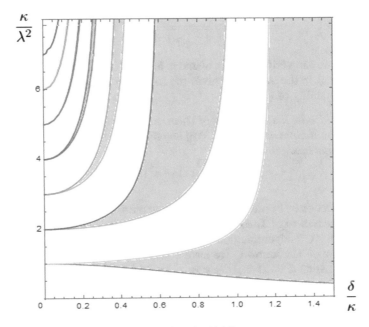

Fig. 4.5 In white, subsets of the stability regions for (4.13)

provided that (4.17) holds, where we have emphasized the relevant variables κ/λ^2 and δ/κ. For the left inequality in (4.22) we use both the bounds $1 < 1 + 2A^2_{\lambda,\kappa} < 3$ and we see that it is certainly satisfied provided that (4.18) holds. $\qquad\square$

The bounds (4.17) and (4.18) give sufficient conditions for the stability and the related regions are depicted in white in Fig. 4.5. It appears that they are "deformations" of the regions in Fig. 3.3. We underline once more that the stability regions may be wider than in Fig. 4.5, although Theorem 4.2 guarantees that they cannot contain horizontal strips, contrary to Fig. 3.3. In fact, at least for $A_{\lambda,\kappa} \approx 0$ (the majority of cases, see Table 4.1), we expect the instability diagram to be very similar to Fig. 3.3, with possibly tiny instability tongues emanating from the points $(0, n)$, for n integer. Formula (4.18) may be used for all integer m but in the case $m = 0$ the estimate can be improved, see [10, Fig. 2]. Assuming that $\kappa < \lambda^2$, we may proceed by using a stability criterion due to Zhukovskii [21] (see also [20, Chap. VIII]), that guarantees linear stability provided that

$$T_\lambda(\delta)^2 \left[\kappa^2 + (1 + 2A^2_{\lambda,\kappa})\delta^2\right] \leqslant \pi^2,$$

where $T_\lambda(\delta)$ is the period of W^2_λ, see (4.15). By using (4.10) and by emphasizing the relevant variables, we obtain the following uniform sufficient condition

$$2\sqrt{2} \int_0^{\pi/2} \frac{d\phi}{\sqrt{2\lambda^4/\kappa^2 + \delta^2/\kappa^2(1 + \sin^2 \phi)}} \sqrt{1 + 3(\delta^2/\kappa^2)} \leqslant \pi.$$

Numerically one can verify that this bound is less restrictive than (4.18). Moreover, by deleting $\sin \phi$ in the integral, it gives the following elegant sufficient condition for the linear stability of (4.13).

Corollary 4.1 *Under the hypotheses of Theorem 4.2, assume moreover that $\kappa < \lambda^2$. Then, the λ-longitudinal mode of (4.13) of amplitude δ is linearly stable with respect to the κ-torsional mode if*

$$\delta^2 \leqslant \frac{2}{5}(\lambda^4 - \kappa^2).$$

We close this section by briefly commenting about the assumption $\kappa^2 \neq \lambda^4$ appearing in the second statement of Theorem 4.2. Such a condition is certainly fulfilled if λ is associated with a C^∞-eigenfunction of (2.8) and κ^2 is a simple eigenvalue of (4.2). Indeed, in view of Theorem 4.1, we have either $\kappa a \in \mathbb{N}$ or $\kappa a \in \mathbb{N} + 1/2$, and in both cases $\kappa(1 - a) \notin \mathbb{N}$. If it were $\kappa = \lambda^2$, writing $\kappa(1 - a) = \lambda(\lambda(1 - a))$ and $\kappa a = \lambda(\lambda a)$ this would be contradicted. Namely,

a simple torsional eigenvalue cannot be a longitudinal eigenvalue associated with a C^∞-eigenfunction.

If λ is associated with a longitudinal eigenfunction which is not C^∞, there are instead cases where $\kappa^2 = \lambda^4$, as Fig. 4.6 shows: precisely, this occurs at each intersection between the bold curves (representing the eigenvalues of (2.8)) and the dashed/dot-dashed ones (representing the eigenvalues of (4.2)); notice that, on the vertical axis, $\sqrt[4]{\mu}$ represents either λ (bold curves) or $\sqrt{\kappa}$ (dashed/dot-dashed curves). However, we checked numerically that the corresponding values of a are not among the ones of our interest, so this possibility, as well as the case when the considered torsional eigenvalue is multiple, will not be further deepened in our study.

4.3.2 Optimal Position of the Piers for Linear and Nonlinear Instability

A full stability analysis of (4.9), for any value of a and any couple of (longitudinal, torsional) modes covers too many different situations. Therefore, we restrict our attention on the most relevant cases, both from a mathematical and a physical point of view, briefly commenting the general situation. The first restriction involves the position of the piers: we take

$$a \geqslant \frac{1}{2}$$

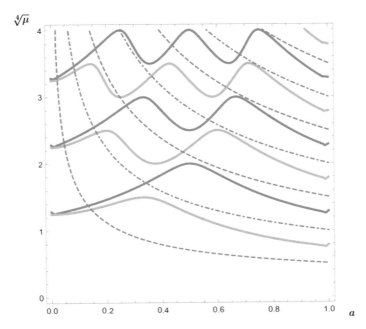

Fig. 4.6 Correspondence between longitudinal (bold) and torsional (dashed/dot-dashed) eigenvalues

because this case contains the physical range (1.1) and, for $a > 1/2$, the second torsional eigenvalue is simple (see the comments at the end of Sect. 4.1). Actually, the instability analysis in case of multiple torsional eigenvalues is much more complicated, due to the strong interactions which occur between the corresponding Fourier components.

The second torsional mode is of crucial importance. Irvine [13, Example 4.6, p. 180] describes the oscillations of the Matukituki Suspension Footbridge as follows: *...the deck persisted in lurching and twisting wildly until failure occurred, and for part of the time a node was noticeable at midspan.* Moreover, according to the detailed analysis on the TNB by Smith–Vincent [19, p. 21], this form of torsional oscillation is the only one ever seen: *The only torsional mode which developed under wind action on the bridge or on the model is that with a single node at the center of the main span.* Therefore,

we restrict our attention to the second torsional mode, (4.24)

that is, we take $\kappa = \kappa_1$. Concerning the longitudinal mode, we take one among the least 12 modes as in Chap. 3, see the motivation in Sect. 3.3. This enables us to exploit several results obtained for the stability of the nonlinear beam Eq. (3.35).

Table 4.2 Energy thresholds of torsional instability for (4.13) with $\kappa = \kappa_1$

a	Linear	Nonlinear	Prevailing mode
0.5	6.276	14.933	e_0
0.56	4.683	14.232	e_0
0.6	3.844	10.276	e_0
2/3	2.783	7.713	e_0
0.7	2.358	7.875	e_0
0.8	1.511	7.616	e_0 (lin) /e_1 (nonlin)
0.9	1.021	3.884	e_0 (lin) /e_1 (nonlin)

With these restrictions, we tackle the problem of the optimal position of the piers and we proceed numerically, analyzing only the energy transfers towards the second torsional mode. For the linear instability, we maintain (3.53) as the characterization of the critical amplitude. We do not expect the picture for linear instability to be significantly different from the one in Fig. 3.3 since $A_{\lambda,\kappa}$ is small, see Table 4.1. On the other hand, the energy threshold of instability is characterized as in Definition 3.5, where nonlinear instability is meant in the sense of Definition 4.2. In Table 4.2 we quote the energy thresholds of linear and nonlinear instability thus found, according to Definitions 4.2 and 4.3.

Some comments are in order. The most important is that, as in the discussion for the beam carried out in Sect. 3.4.5,

linear instability is a clue for the occurrence of nonlinear instability.

We also notice that instability appears first for low prevailing longitudinal modes, since the ratio κ^2/λ^4 becomes very small on growing of λ; hence, looking at both Fig. 3.3 (for $A_{\lambda,\kappa} = 0$) and Fig. 4.5 (for $A_{\lambda,\kappa} \neq 0$), we understand that the more the ratio κ^2/λ^4 approaches 0, the larger is the energy needed to reach instability. In some cases instability is never reached, this occurring for instance when $A_{\lambda,\kappa} = 0$ and $\lambda^2 > \kappa$, since in this case we fall into the first linear stability region in Fig. 3.3 (the one below $\rho^4/\lambda^4 = 1$).

In the next section, we will analyze the role of a further nonlinearity.

4.4 The Case of Extensible Hangers

4.4.1 Choice of the Nonlinearity

Not always the hangers may be considered rigid, which means that they can lose tension if the deck and the cables are too close to each other. This phenomenon is called *slackening* and was observed during the TNB collapse, see [1, V–12]. The

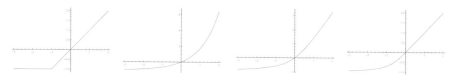

Fig. 4.7 Graphs of the functions in (4.25)

restoring action of the hangers is described by a *local nonlinearity* that should be able to represent a slightly superlinear spring when extended and only gravity when slacken. Recalling the convention (1.4), several nonlinearities of this kind have been considered in literature: among others, we mention here

$$s \mapsto (1+s)^+ - 1, \quad s \mapsto e^s - 1, \quad s \mapsto s - 1 + \sqrt{1+s^2}, \quad s \mapsto \begin{cases} \frac{s}{\sqrt{1+s^2}} & \text{if } s \leqslant 0 \\ s & \text{if } s \geqslant 0 \end{cases},$$
(4.25)

whose graphs are represented in the pictures of Fig. 4.7: one can notice that their qualitative behavior is the same.

The first nonlinearity in (4.25) was introduced by Lazer–McKenna [14] and describes a linear restoring force when the hangers are in tension and a constant gravity force (normalized to be -1) when they slacken. This nonlinearity, that well describes the slackening mechanism, has two drawbacks. First, it is nonsmooth and this makes the numerical experiments very difficult. Second, Brownjohn [5, p. 1364] remarks that *the hangers are critical elements in a suspension bridge and for large-amplitude motion their behaviour is not well modelled by either simple on/off stiffness or invariant connections.* Moreover, supported by analytical and experimental studies on the dynamic response of suspension bridges, Brownjohn [5] is able to show a strong nonlinear contribution of the cable/hanger effects while McKenna–Tuama [18] write *...we expect the bridge to behave like a stiff spring, with a restoring force that becomes somewhat superlinear.* For all these reasons, one is led to slightly modify the first nonlinearity in (4.25). The second, third, fourth nonlinearities have been considered, respectively, in [3, 15, 17], as possible smooth alternative choices for the restoring force. They all tend to -1 as $s \to -\infty$ and they all have first derivative equal to 1 when $s = 0$: this number represents the *Hooke constant of elasticity* of the hangers. Since we aim at emphasizing the role of the elasticity of the hangers, we adopt here the modified nonlinearity

$$f_\varsigma(s) = \varsigma s - 1 + \sqrt{1 + \varsigma^2 s^2} \qquad (\varsigma > 0)$$

which is a variant of the third nonlinearity in (4.25). We still have $f_\varsigma(s) \to -1$ as $s \to -\infty$ so that the gravity constant -1 is conserved. Moreover, $f'_\varsigma(0) = \varsigma$ so that ς measures the elasticity of the hangers. Finally, $f_\varsigma(s) \sim 2\varsigma s$ as $s \to +\infty$, showing that $f_\varsigma(s)$ has a slightly superlinear behavior for $s > 0$, with a slope going from ς

towards 2ς. Note also that $f_\varsigma \to 0$ in $L^\infty_{\text{loc}}(\mathbb{R})$ for $\varsigma \to 0$, so that for vanishing ς we are back in the situation of rigid hangers, see Sect. 4.3.

By taking $f = f_\varsigma$ for some $\varsigma > 0$, system (4.4) becomes

$$
\begin{cases}
u_{tt} + u_{xxxx} + \left(\int_I (u^2 + \theta^2)\right)u + 2\left(\int_I u\theta\right)\theta + 2\varsigma u + \left[\sqrt{1 + \varsigma^2(u+\theta)^2} + \sqrt{1 + \varsigma^2(u-\theta)^2} - 2\right] = 0 \\
\theta_{tt} - \theta_{xx} + 2\left(\int_I u\theta\right)u + \left(\int_I (u^2 + \theta^2)\right)\theta + 2\varsigma\theta + \left[\sqrt{1 + \varsigma^2(u+\theta)^2} - \sqrt{1 + \varsigma^2(u-\theta)^2}\right] = 0.
\end{cases}
\tag{4.26}
$$

We complement (4.26) with the boundary-internal-initial conditions (4.5)–(4.6) and we define weak solutions as in Definition 4.1. Since f_ς satisfies the two latter conditions in (3.3), Proposition 4.1 can be applied and problem (4.26) is well-posed.

4.4.2 Linear and Nonlinear Instability for the Two Modes System

As already mentioned in Sect. 4.2, the spaces $V(I) \times \{0\}$ and $\{0\} \times W(I)$ are invariant for (4.26): if we take $(\theta_0, \theta_1) = (0, 0)$, then the weak solution (u, θ) of (4.26)–(4.6) has no torsional component ($\theta \equiv 0$) while the longitudinal component u satisfies the equation

$$
u_{tt} + u_{xxxx} + \left(\int_I u^2\right)u + 2\left(\varsigma u - 1 + \sqrt{1 + \varsigma^2 u^2}\right) = 0 \quad x \in I, \quad t > 0,
$$

together with the corresponding conditions in (4.5). Also this problem should be intended in its weak form, that is,

$$
\langle u_{tt}, v \rangle_V + \int_I u_{xx}v'' + \int_I u^2 \int_I uv + 2\int_I \left(\varsigma u - 1 + \sqrt{1 + \varsigma^2 u^2}\right)v = 0 \quad \forall v \in V(I), \ t > 0.
$$

As in Sect. 4.3, we reduce our analysis to the case of approximate two-modes solutions of (4.26) in the form (4.12). To this end, for any couple (λ^4, κ^2) of eigenvalues of (2.8) and (4.2) we need to introduce the two functions

$$
\Gamma^{\lambda,\kappa}_\varsigma(w, z) := \int_I \left[\sqrt{1 + \varsigma^2\left(we_\lambda(x) + z\eta_\kappa(x)\right)^2} + \sqrt{1 + \varsigma^2\left(we_\lambda(x) - z\eta_\kappa(x)\right)^2} - 2\right]e_\lambda(x)\,dx,
$$

$$
\Xi^{\lambda,\kappa}_\varsigma(w, z) := \int_I \left[\sqrt{1 + \varsigma^2\left(we_\lambda(x) + z\eta_\kappa(x)\right)^2} - \sqrt{1 + \varsigma^2\left(we_\lambda(x) - z\eta_\kappa(x)\right)^2}\right]\eta_\kappa(x)\,dx.
$$

These functions have the following symmetries:

$$
\begin{aligned}
\Gamma^{\lambda,\kappa}_\varsigma(w, z) &= \Gamma^{\lambda,\kappa}_\varsigma(-w, z) = \Gamma^{\lambda,\kappa}_\varsigma(w, -z) = \Gamma^{\lambda,\kappa}_\varsigma(-w, -z) \\
\Xi^{\lambda,\kappa}_\varsigma(w, z) &= -\Xi^{\lambda,\kappa}_\varsigma(-w, z) = -\Xi^{\lambda,\kappa}_\varsigma(w, -z) = \Xi^{\lambda,\kappa}_\varsigma(-w, -z)
\end{aligned}
\quad \forall(w, z) \in \mathbb{R}^2.
\tag{4.27}
$$

Then the time-dependent coefficients w, z of approximate two-modes solutions of (4.26) in the form (4.12) satisfy the system

$$\begin{cases} \ddot{w}(t) + (\lambda^4 + 2\varsigma)w(t) + (1 + 2A_{\lambda,\kappa}^2)z(t)^2 w(t) + w(t)^3 + \Gamma_\varsigma^{\lambda,\kappa}(w(t), z(t)) = 0 \\ \ddot{z}(t) + (\kappa^2 + 2\varsigma)z(t) + (1 + 2A_{\lambda,\kappa}^2)w(t)^2 z(t) + z(t)^3 + \Xi_\varsigma^{\lambda,\kappa}(w(t), z(t)) = 0, \end{cases}$$
(4.28)

to which we associate the initial conditions

$$w(0) = \delta, \quad z(0) = z_0, \quad \dot{w}(0) = \dot{z}(0) = 0. \tag{4.29}$$

If $z_0 = 0$, then the solution of (4.28)–(4.29) is $(w, z) = (W_\lambda, 0)$, where W_λ solves

$$\ddot{W}_\lambda(t) + (\lambda^4 + 2\varsigma)W_\lambda(t) + W_\lambda(t)^3 + \Psi_\varsigma^\lambda(W_\lambda(t)) = 0, \quad W_\lambda(0) = \delta, \quad \dot{W}_\lambda(0) = 0, \tag{4.30}$$

with

$$\Psi_\varsigma^\lambda(w) = 2 \int_I \left[\sqrt{1 + \varsigma^2 w^2 e_\lambda(x)^2} - 1 \right] e_\lambda(x)\, dx.$$

The following statement holds.

Proposition 4.3 *Let $g(s) := (\lambda^4 + 2\varsigma)s + s^3 + \Psi_\varsigma^\lambda(s)$ and $G(s) := \int_0^s g(\tau)d\tau$. Then, for any $\delta > 0$, the solution W_λ of (4.30) is periodic with period*

$$\tau_\lambda = \sqrt{2} \int_{\gamma_\delta}^\delta \frac{ds}{\sqrt{G(\delta) - G(s)}}, \tag{4.31}$$

where γ_δ is the unique negative solution of $G(\gamma_\delta) = G(\delta)$.

To see this, it suffices to notice that

$$\left| \frac{d}{ds} \Psi_\varsigma^\lambda(s) \right| = 2\varsigma \left| \int_I \frac{\varsigma s e_\lambda(x)^3}{\sqrt{1 + \varsigma^2 s^2 e_\lambda(x)^2}}\, dx \right| \leqslant 2\varsigma \int_I e_\lambda(x)^2 dx = 2\varsigma,$$

which shows that g is strictly increasing over $\mathbb{R}$ and that G is strictly convex. Therefore, the sublevels of the energy

$$\frac{\dot{w}^2}{2} + G(w)$$

are convex sets of the phase plane $(w, \dot{w})$ and its level lines are closed curves. This shows that any solution of (4.30) is periodic; a standard computation then provides expression (4.31).

In fact, the "unpleasant parts" in (4.28) vanish whenever the longitudinal eigenfunction is odd.

Proposition 4.4 *Assume that e_λ is an odd eigenfunction of (2.8), as given by Theorem 2.3. Then $\Gamma_\varsigma^{\lambda,\kappa}(w, z) = \Xi_\varsigma^{\lambda,\kappa}(w, z) = 0$ for any $(w, z) \in \mathbb{R}^2$ and any eigenfunction η_κ of (4.2), as given by Theorem 4.1.*

To prove Proposition 4.4 one needs to distinguish two cases. If η_κ is also odd, then $\left(we_\lambda(x) \pm z\eta_\kappa(x)\right)^2$ are even functions so that the integrands in $\Gamma_\varsigma^{\lambda,\kappa}$ and $\Xi_\varsigma^{\lambda,\kappa}$ are odd and the integrals vanish. If η_κ is even, then one combines (4.27) with the change of variables $x \mapsto -x$ within the integrals that define $\Gamma_\varsigma^{\lambda,\kappa}$ and $\Xi_\varsigma^{\lambda,\kappa}$.

Similarly as in Definition 4.3, we characterize the linear stability of longitudinal modes as follows.

Definition 4.4 The mode W_λ is said to be *linearly stable* (*unstable*) with respect to the κ-torsional-mode if $\xi \equiv 0$ is a stable (unstable) solution of the linear Hill equation

$$\ddot{\xi}(t) + \left(\kappa^2 + 2\varsigma + (1 + 2A_{\lambda,\kappa}^2)W_\lambda(t)^2 + 2\varsigma^2 B_\varsigma(W_\lambda(t))\right)\xi(t) = 0, \qquad (4.32)$$

where

$$B_\varsigma(w) = \int_I \frac{e_\lambda(x)\eta_\kappa(x)^2 w}{\sqrt{1 + \varsigma^2 w^2 e_\lambda(x)^2}} \, dx.$$

Equation (4.32) is obtained by linearizing the second equation in system (4.28) around the solution $(w, z) = (W_\lambda, 0)$, obtained with initial datum $z_0 = 0$. It is a Hill equation, since the coefficient multiplying ξ is periodic, see Proposition 4.3.

We now state a result which highlights striking differences between odd and even longitudinal modes when slackening occurs; in particular, it will allow us to simplify (4.32) in some cases.

Theorem 4.3 *Let $\varsigma \neq 0$ and let λ^4 and κ^2 be, respectively, eigenvalues of (2.8) and (4.2), with associated eigenfunctions e_λ and η_κ provided by Theorems 2.3 and 4.1, respectively. Then:*

- *if e_λ is odd, then $B_\varsigma(w) = 0$ for all $w \in \mathbb{R}$;*
- *if e_λ is even and either $\eta_\kappa = D_\kappa$ or $\eta_\kappa = P_\kappa$ (being zero on the side spans), then $B_\varsigma(w) \not\equiv 0$ in any neighborhood of $w = 0$.*

Proof If e_λ is odd, the statement follows easily from the fact that, if η_κ is an eigenfunction of (4.2) as in Theorem 4.1, then η_κ^2 is even and the integrand defining B_ς is odd whenever e_λ is odd.

The difficult part is the second item, which we now prove. According to Theorem 2.3, if e_λ is even we need to distinguish two cases.

- **Case (I):** $e_\lambda(x) = E_\lambda(x) = \cos(\lambda x)$.

In this case, the following calculus formula turns out to be useful:

$$\lambda \neq 2\kappa \implies \int_0^{a\pi} \cos(\lambda x)\cos(2\kappa x)\,dx = \frac{\lambda\sin(\lambda a\pi)\cos(2\kappa a\pi) - 2\kappa\cos(\lambda a\pi)\sin(2\kappa a\pi)}{\lambda^2 - 4\kappa^2}.$$

In particular, since $\lambda a - 1/2 \in \mathbb{N}$ (see Theorem 2.3), this yields $\cos(\lambda a\pi) = 0$ and hence

$$\lambda \neq 2\kappa \implies \int_0^{a\pi} \cos(\lambda x) \cos(2\kappa x) \, dx = \frac{\lambda \sin(\lambda a\pi) \cos(2\kappa a\pi)}{\lambda^2 - 4\kappa^2}. \quad (4.33)$$

Since $B_\varsigma(w) = B_1(\varsigma w)/\varsigma$, it suffices to prove that $B_1(w) \not\equiv 0$ in any neighborhood of $w = 0$. And since $B_1(w) = w \int_I e_\lambda \eta_\kappa^2 + o(w^2)$ as $w \to 0$, in order to prove this fact it suffices to show that

$$J := \int_I e_\lambda(x) \eta_\kappa(x)^2 \, dx = 2 \int_0^{a\pi} e_\lambda(x) \eta_\kappa(x)^2 \, dx \neq 0. \quad (4.34)$$

Theorem 4.1 suggests to distinguish two further subcases:

$$\text{(Ia)} \quad \eta_\kappa(x) = \chi_0(x) \sin(\kappa x) \qquad \text{(Ib)} \quad \eta_\kappa(x) = \chi_0(x) \cos(\kappa x).$$

In Case (Ia) we have $\sin(\kappa a\pi) = 0$ and therefore

$$\kappa a \in \mathbb{N}, \qquad \cos(2\kappa a\pi) = 1. \quad (4.35)$$

From Theorem 2.3 and (4.35) we know that there exist two integers m and n such that $\lambda = (2m + 1)/2a$ and $\kappa = n/a$. Therefore,

$$4\kappa^2 - \lambda^2 = \frac{16n^2 - 4m^2 - 4m - 1}{4a^2} \neq 0$$

since the first three terms in the numerator are multiples of 4 while -1 is not. This shows that (4.33) applies and, by (4.33) and using again (4.35), we obtain

$$\begin{aligned} J &= 2 \int_0^{a\pi} \cos(\lambda x) \sin^2(\kappa x) dx = \int_0^{a\pi} \cos(\lambda x)[1 - \cos(2\kappa x)] dx \\ &= \frac{\sin(\lambda a\pi)}{\lambda} - \frac{\lambda \sin(\lambda a\pi)}{\lambda^2 - 4\kappa^2} = \frac{4\kappa^2 \sin(\lambda a\pi)}{\lambda(4\kappa^2 - \lambda^2)}. \end{aligned}$$

Since $\sin(\lambda a\pi) = \pm 1$, this proves (4.34) in Case (Ia).

In Case (Ib) we have $\cos(\kappa a\pi) = 0$ and therefore

$$\kappa a - \frac{1}{2} \in \mathbb{N}, \qquad \cos(2\kappa a\pi) = -1. \quad (4.36)$$

From Theorem 2.3 and (4.36) we know that there exist two integers m and n such that $\lambda = (2m + 1)/2a$ and $\kappa = (2n + 1)/2a$. Therefore,

$$4\kappa^2 - \lambda^2 = \frac{16n^2 + 16n - 4m^2 - 4m + 3}{4a^2} \neq 0$$

since the first four terms in the numerator are multiples of 4 while 3 is not. This shows that (4.33) applies and, by (4.33) and using again (4.36), we obtain

$$J = 2 \int_0^{a\pi} \cos(\lambda x) \cos^2(\kappa x) dx = \int_0^{a\pi} \cos(\lambda x)[1 + \cos(2\kappa x)] dx$$
$$= \frac{\sin(\lambda a\pi)}{\lambda} - \frac{\lambda \sin(\lambda a\pi)}{\lambda^2 - 4\kappa^2} = \frac{4\kappa^2 \sin(\lambda a\pi)}{\lambda(4\kappa^2 - \lambda^2)}.$$

Since $\sin(\lambda a\pi) = \pm 1$, this proves (4.34) also in Case (Ib).

This completes the proof of Theorem 4.3 for e_λ even in Case (I).

- **Case (II):** $e_\lambda(x) = \mathscr{E}_\lambda(x) = C[\cos(\lambda x) - \frac{\cos(\lambda a\pi)}{\cosh(\lambda a\pi)} \cosh(\lambda x)]$ if $x \in [0, a\pi]$.

In this case, we start with a technical statement that probably holds in a more general form, but which is strong enough for our purposes.

Lemma 4.2 *Assume that $0 < |A| < 1$ and consider the function $f(t) = \cos t + A \cosh t$. Then, f admits a finite number of critical points (local extrema) in $[0, \infty)$, say $\{t_1, ..., t_m\}$ for some integer $m \geqslant 1$. Moreover, f cannot have a maximum t_i at positive level and a minimum t_j at negative level such that $|f(t_j)| = f(t_i)$.*

Proof The first statement follows by noticing that $f'(t) = -\sin t + A \sinh t$ so that $f'(0) = 0$ (proving $m \geqslant 1$) and f' has eventually the same sign as A. As for the second statement, we first claim that

$$\text{if } |f(t_i)| = |f(t_j)| \text{ for some } i \neq j, \text{ then } \cos t_i \cos t_j > 0. \tag{4.37}$$

To see this, put together the three facts

$$f(t_i)^2 = f(t_j)^2, \quad f'(t_i) = 0 \Leftrightarrow \sin t_i = A \sinh t_i, \quad f'(t_j) = 0 \Leftrightarrow \sin t_j = A \sinh t_j,$$

to obtain $\cos t_i \cosh t_i = \cos t_j \cosh t_j$, which shows that either $\cos t_i = \cos t_j = 0$ or $\cos t_i \cos t_j > 0$. The claim follows if we exclude the first possibility; by contradiction, if it were true then $|\sin t_i| = |\sin t_j| = 1$ and, since $f'(t_i) = f'(t_j) = 0$, this would imply that $\sinh t_i = \sinh t_j$, contradicting $t_i \neq t_j$.

Assume now by contradiction that f has a maximum t_i at positive level and a minimum t_j at negative level satisfying $|f(t_j)| = f(t_i)$. Then, from $f(t_i)f(t_j) < 0$ we infer that

$$0 > \cos t_i \cos t_j + A^2 \cosh t_i \cosh t_j + A \cos t_i \cosh t_j + A \cos t_j \cosh t_i$$
$$> A(\cos t_i \cosh t_j + \cos t_j \cosh t_i), \tag{4.38}$$

where we used (4.37). On the other hand, from $f''(t_i)f''(t_j) \leqslant 0$ and using again (4.37), we infer that

$$0 \geqslant \cos t_i \cos t_j + A^2 \cosh t_i \cosh t_j - A \cos t_i \cosh t_j - A \cos t_j \cosh t_i$$
$$> -A(\cos t_i \cosh t_j + \cos t_j \cosh t_i),$$

which contradicts (4.38), concluding the proof. □

We now continue the analysis of Case (II). Consider the function

$$A(w) := \frac{B_1(w)}{w} = \int_0^{a\pi} \frac{e_\lambda(x)\eta_\kappa(x)^2}{\sqrt{1+w^2 e_\lambda(x)^2}} \, dx \quad \forall w \neq 0, \qquad A(0) = \int_0^{a\pi} e_\lambda(x)\eta_\kappa(x)^2 \, dx,$$

so that Theorem 4.3 will be proved if we show that $A(w) \not\equiv 0$ in any neighborhood of $w = 0$. Assume by contradiction this to be false, namely

$$A(w) \equiv 0 \text{ in some neighborhood } U \text{ of } w = 0 \implies A^{(k)}(w) \equiv 0 \text{ in } U \; \forall k \in \mathbb{N}, \tag{4.39}$$

where $A^{(k)}(w)$ denotes the k-th derivative of A with respect to w. We have

$$A'(w) = -w \int_0^{a\pi} \frac{e_\lambda(x)^3 \eta_\kappa(x)^2}{[1+w^2 e_\lambda(x)^2]^{3/2}} \, dx$$

and by (4.39) we deduce that

$$A_1(w) := \int_0^{a\pi} \frac{e_\lambda(x)^3 \eta_\kappa(x)^2}{[1+w^2 e_\lambda(x)^2]^{3/2}} \, dx \equiv 0 \text{ in } U, \qquad A_1(0) = \int_0^{a\pi} e_\lambda(x)^3 \eta_\kappa(x)^2 \, dx = 0.$$

In turn, by differentiating $A_1(w)$, we deduce that

$$A_2(w) := \int_0^{a\pi} \frac{e_\lambda(x)^5 \eta_\kappa(x)^2}{[1+w^2 e_\lambda(x)^2]^{5/2}} \, dx \equiv 0 \text{ in } U, \qquad A_2(0) = \int_0^{a\pi} e_\lambda(x)^5 \eta_\kappa(x)^2 \, dx = 0.$$

By iterating this procedure, we obtain that

$$\int_0^{a\pi} e_\lambda(x)^{2k+1} \eta_\kappa(x)^2 \, dx = 0 \qquad \forall k \in \mathbb{N}. \tag{4.40}$$

By Lemma 4.2 we may find $K \neq 0$ such that $g_\lambda(x) := K e_\lambda(x)$ satisfies

$$\exists \bar{x} \in [0, a\pi) \quad \text{s.t.} \quad g_\lambda(\bar{x}) = \|g\|_{L^\infty(0, a\pi)} > 1, \qquad g_\lambda(x) \geq -1 \; \forall x \in [0, a\pi).$$

Lemma 4.2 does not specify if such $\bar{x}$ is unique, nevertheless we may consider the nonempty open set H where $g_\lambda(x) > 1$: this may be an interval (in case of uniqueness of $\bar{x}$) or the union of a finite number of intervals (one for each $\bar{x}$). From (4.40) we infer that

$$\int_0^{a\pi} g_\lambda(x)^{2k+1} \eta_\kappa(x)^2 \, dx = 0 \qquad \forall k \in \mathbb{N}.$$

But, denoting $H_0 = (0, a\pi) \backslash H$, we have that for every $k \in \mathbb{N}$ it holds

$$\int_0^{a\pi} g_\lambda(x)^{2k+1} \eta_\kappa(x)^2 \, dx = \int_{H_0} g_\lambda(x)^{2k+1} \eta_\kappa(x)^2 \, dx + \int_H g_\lambda(x)^{2k+1} \eta_\kappa(x)^2 \, dx = 0.$$

By letting $k \to \infty$, the first term converges to 0 by the Lebesgue Theorem (recall that $|g_\lambda(x)| < 1$ a.e. in H_0) while the second term diverges to $+\infty$ (recall that $g_\lambda(x) > 1$ in H). This gives a contradiction.

Therefore, (4.39) does not hold and this proves Theorem 4.3 for even eigenfunctions e_λ also in Case (II). $\square$

Theorem 4.3 does not clarify what happens in case of a multiple torsional eigenvalue κ, when choosing the associated eigenfunctions differently from Theorem 4.1. Also the case when e_λ is even and η_κ is an eigenfunction of (4.2) other than $\mathbf{D}_\kappa$ or $\mathbf{P}_\kappa$ is not considered. We refer the reader to the discussion in Sect. 5.2.

Combining Theorem 4.3 with (4.11), we can simplify (4.32).

Corollary 4.2 Let λ^4 and κ^2 be, respectively, eigenvalues of (2.8) and (4.2), with associated eigenfunctions e_λ and η_κ provided by Theorems 2.3 and 4.1, respectively. Then:

- if e_λ is odd and η_κ is even, then Eq. (4.32) simplifies to

$$\ddot{\xi}(t) + \left(\kappa^2 + 2\varsigma + W_\lambda(t)^2\right)\xi(t) = 0 \, ;$$

- if e_λ is odd and η_κ is odd, then (4.32) simplifies to

$$\ddot{\xi}(t) + \left(\kappa^2 + 2\varsigma + (1 + 2A_{\lambda,\kappa}^2)W_\lambda(t)^2\right)\xi(t) = 0 \, ;$$

- if e_λ is even and η_κ is odd, then (4.32) simplifies to

$$\ddot{\xi}(t) + \left(\kappa^2 + 2\varsigma + W_\lambda(t)^2 + 2\varsigma^2 B_\varsigma(W_\lambda(t))\right)\xi(t) = 0.$$

For the linear stability of system (4.28), we thus have to take into account all these possible combinations, resulting in the following theorem.

Theorem 4.4 Let $\varsigma \neq 0$ and let λ^4 and κ^2 be, respectively, eigenvalues of (2.8) and (4.2), with κ^2 associated with an eigenfunction of the kind $\mathbf{D}_\kappa$ or $\mathbf{P}_\kappa$ (being zero on the side spans). The following hold:

- the λ-longitudinal-mode of (4.28) of amplitude δ is linearly unstable with respect to the κ-torsional-mode whenever δ is sufficiently large;
- if e_λ is odd, then the λ-longitudinal-mode of (4.28) of amplitude δ is linearly stable with respect to the κ-torsional-mode if δ is sufficiently small, provided that $(\kappa^2 + 2\varsigma)/(\lambda^4 + 2\varsigma) \notin \mathbb{N}^2$;
- if e_λ is even, then the λ-longitudinal-mode of (4.28) of amplitude δ is linearly stable with respect to the κ-torsional-mode if δ is sufficiently small, provided that $4(\kappa^2 + 2\varsigma)/(\lambda^4 + 2\varsigma) \notin \mathbb{N}^2$.

Proof In order to prove instability for large δ, we proceed as in Lemma 4.1, with a few changes. For all $\varepsilon > 0$ we define Z_ε and ξ_ε by

$$W_\lambda(t) = \sqrt{\frac{\lambda^4 + 2\varsigma}{\varepsilon}} \, Z_\varepsilon\left(\sqrt{\frac{\lambda^4 + 2\varsigma}{\varepsilon}}\, t\right), \qquad \xi(t) = \xi_\varepsilon\left(\sqrt{\frac{\lambda^4 + 2\varsigma}{\varepsilon}}\, t\right).$$

If W_λ satisfies (4.30), then Z_ε solves the equation

$$\ddot{Z}_\varepsilon(t) + \varepsilon \dot{Z}_\varepsilon(t) + Z_\varepsilon(t)^3 + \frac{2\varepsilon}{\lambda^4 + 2\varsigma} \int_I \left(\sqrt{\frac{\varepsilon}{\lambda^4 + 2\varsigma} + \varsigma^2 Z_\varepsilon(t)^2 e_\lambda(x)^2} - \sqrt{\frac{\varepsilon}{\lambda^4 + 2\varsigma}} \right) e_\lambda(x)\,dx = 0$$
$$(4.41)$$

and, according to Definition 4.4, the linear stability of (4.28) depends on the stability of the following Hill equation:

$$\ddot{\xi}_\varepsilon(t) + \left[\frac{(\kappa^2 + 2\varsigma)\varepsilon}{\lambda^4 + 2\varsigma} + (1 + 2A_{\lambda,\kappa}^2) Z_\varepsilon(t)^2 + \frac{2\varepsilon\varsigma^2 Z_\varepsilon(t)}{\sqrt{\lambda^4 + 2\varsigma}} \int_I \frac{e_\lambda(x)\eta_\kappa(x)^2\,dx}{\sqrt{\varepsilon + \varsigma^2(\lambda^4 + 2\varsigma) Z_\varepsilon(t)^2 e_\lambda(x)^2}} \right] \xi_\varepsilon(t) = 0.$$
$$(4.42)$$

By letting $\varepsilon \to 0$ we see that (4.41) and (4.42) "converge" again to the limit problems (4.21) and we conclude as for Lemma 4.1, by invoking [6, Theorem 3.1].

To finish the proof of Theorem 4.4, we apply the Burdina criterion (Lemma 3.1) to the periodic coefficient of the Hill equation (4.32), given by $p(t) = \kappa^2 + 2\varsigma + (1 + 2A_{\lambda,\kappa}^2) W_\lambda^2(t) + 2\varsigma^2 B_\varsigma(W_\lambda(t))$. By Theorem 4.3, we know that $B_\varsigma(W_\lambda(t)) \not\equiv 0$ if and only if e_λ is even. Thus, $p(t)$ has period equal to $\tau_\lambda/2$ if e_λ is odd, and τ_λ if e_λ is even (note that such expressions depend on δ), where τ_λ is as in (4.31). For $\delta \to 0$ these periods converge, respectively, to $2\pi/\sqrt{\lambda^4 + 2\varsigma}$ and to $\pi/\sqrt{\lambda^4 + 2\varsigma}$ while $p(t)$ converges to $\kappa^2 + 2\varsigma$. Hence, recalling (3.49), for $\delta \to 0$ we are in a stability region if there exists $k \in \mathbb{N}$ such that

$$k\pi < \frac{2\pi}{\sqrt{\lambda^4 + 2\varsigma}} \sqrt{\kappa^2 + 2\varsigma} < (k+1)\pi \qquad \text{if } e_\lambda \text{ is even},$$

and

$$k\pi < \frac{\pi}{\sqrt{\lambda^4 + 2\varsigma}} \sqrt{\kappa^2 + 2\varsigma} < (k+1)\pi \qquad \text{if } e_\lambda \text{ is odd},$$

from which the statement for small δ follows. $\square$

This statement should be compared with Theorem 4.2. Again, it merely gives qualitative sufficient conditions both for linear stability and instability. Some differences appear here due to the presence of the term B_ς, which makes the coefficient of ξ in (4.32) τ_λ (and not $\tau_\lambda/2$)-periodic, thus causing a discontinuity with respect to the case $\varsigma = 0$. This shows that in general any (even small) nonlinearity due to the hangers may destroy the stability diagram for a given system studied without hangers.

More results are obtained numerically, as reported in the next section. Therein we also study numerically the nonlinear stability (according to Definition 4.2) and we seek the optimal position of the piers that maximizes the critical energy.

4.4.3 Optimal Position of the Piers in Degenerate Plates

In this section, we analyze the stability of the second torsional mode (cf. (4.24)) for system (4.28). The procedure is similar to the one in Sect. 4.3.2: however, we have already seen that the presence of f_ς makes the situation richer, alternatively involving each of the equations displayed in Corollary 4.2.

The situation is already more complicated at the level of linear stability, since we do not have a precise picture of the resonance tongues for equation (4.32), especially if e_λ is even. In principle, it may happen to cross thick instability regions which alternate to stability ones; hence the definition of critical amplitude given in (3.53) would be nonsense. Nevertheless, we could not numerically detect any of them. We thus simply took as critical amplitude threshold of linear instability the least value of δ for which the absolute value of the monodromy matrix associated with (4.32) is larger than 2. Concerning nonlinear instability, we proceeded as usual, following the steps in Sect. 4.5.

As for the choice of ς, we are guided by the claims reported in the engineering literature [2, 16] (see Sect. 1): the cables represent the main source of nonlinearity, so we stick to *small values of* ς. A full quantitative analysis would require a huge effort and would be useless without taking into account the exact values of the real parameters inside the considered equations, see Sect. 5.2 for further comments. We chose to perform our analysis for $\varsigma \in \{0.1, 0.2, 0.5, 1\}$. The corresponding results are displayed in Table 4.3. For the reader's convenience, we also report the results for $\varsigma = 0$. The global energy of the system, given by

$$E_\varsigma(u, \theta) = \frac{1}{2}\int_I (u_t^2 + \theta_t^2 + \theta_x^2 + u_{xx}^2) + \frac{1}{8}\left[\int_I (u+\theta)^2\right]^2 + \frac{1}{8}\left[\int_I (u-\theta)^2\right]^2$$
$$+ \int_I F_\varsigma(u+\theta) + \int_I F_\varsigma(u-\theta)$$

with

$$F_\varsigma(s) = \frac{1}{2}\left(\varsigma s^2 - 2s + s\sqrt{1+\varsigma^2 s^2} + \frac{\operatorname{arcsinh}(\varsigma s)}{\varsigma}\right),$$

increases with the hangers elasticity ς, thus E_ς is not the most suitable parameter for the comparison of the stability performances on varying of ς. Much more meaningful appears the critical amplitude. For this reason, in Table 4.3 we report the two critical amplitudes δ_{lin} and δ for linear and nonlinear stability, respectively. Table 4.3 also contains the time τ in correspondence of which nonlinear instability is observed and the associated expansion rate $\mathcal{ER}_\tau$, which is around 100, in line with what we observed for the beam.

It appears evident that, for a given a, $\varsigma \mapsto \delta_{lin}(\varsigma)$ is decreasing, showing that increasing the elasticity of the hangers lowers the linear stability. The pattern for nonlinear instability is by far less clear. First, for $a = 0.8$ and $a = 0.9$ the longitudinal mode having the smallest instability threshold is e_1 (odd), while for the other values of

Table 4.3 Linear and nonlinear instability for (4.28)

a	ς	δ_{lin}	δ	τ	$\mathcal{ER}_\tau$
0.5	0	1.77	2.38	15.61	70.07
	0.1	1.77	2.44	15.16	69.85
	0.2	1.77	2.51	14.68	69.45
	0.5	1.75	2.54	15.96	97.78
	1	1.7	2.64	14.76	74.14
0.56	0	1.71	2.45	15.95	125.61
	0.1	1.71	2.4	15.95	101.08
	0.2	1.7	2.35	15.96	83.37
	0.5	1.68	2.41	15.08	67.43
	1	1.63	2.43	15.96	77.84
0.6	0	1.66	2.27	14.89	65.64
	0.1	1.66	2.38	14.17	65.59
	0.2	1.65	2.42	15.99	116
	0.5	1.63	2.28	15.94	68.05
	1	1.58	2.49	15.97	95.98
2/3	0	1.57	2.15	15.97	78.6
	0.1	1.57	2.13	15.89	66.21
	0.2	1.56	2.24	15.03	65.92
	0.5	1.54	2.35	15.96	93.56
	1	1.49	2.49	14.31	61.77
0.7	0	1.52	2.19	15.94	95.91
	0.1	1.52	2.14	15.94	75.82
	0.2	1.52	2.14	15.73	66.1
	0.5	1.5	2.38	15.99	103.57
	1	1.45	2.44	14.62	61.53
0.8	0	1.39	1.6*	15.71	100.07
	0.1	1.39	1.61*	15.44	99.19
	0.2	1.39	1.61*	15.28	88.08
	0.5	1.37	1.64*	14.56	85.26
	1	1.32	1.64*	15.82	92.27
0.9	0	1.28	1.36*	15.8	81.91
	0.1	1.27	1.38*	15.36	82.02
	0.2	1.27	1.4*	14.97	81.62
	0.5	1.25	1.45*	14.05	74.36
	1	1.21	1.46*	15.33	82.52

a it is e_0 (even). Hence, one should not compare the critical amplitudes for $a = 0.8$, $a = 0.9$ (marked with the symbol $*$ in Table 4.3 for this reason) with the others; actually, they are much smaller. Moreover, they are *increasing* with respect to ς, a feature observed in all our experiments with odd longitudinal modes. As for the other values of a, there is no clear emerging pattern; hence, in this case linear stability gives important complementary information. Overall, we observe that

> **in the range** (1.1),**the best placement of the piers for system** (4.28)
> **lies around** $a = 0.5$.

4.5 The Algorithm for the Computation of the Energy Thresholds

In this section, we briefly describe the program we implemented for the numerical simulations in order to check whether nonlinear instability appears. The program has been written and run using Mathematica© software. Recalling Definition 3.4, the aim is to determine whether a residual mode has increased its oscillations so as to subvert of an order of magnitude the inequality in (3.10) and it has grown sufficiently abruptly on itself. This is done in an iterative way for different values of $a \in (0, 1)$, as follows.

- *Step 1.* Fix $0 < a < 1$, $N \in \{2, 3, 4, \ldots\}$, $0 < \eta < 1$, $0 < tol \ll 1$, $0 < step \ll 1$, $0 < r < 1$, $EN_{\max} > 0$ and set $j = 0$.
- *Step 2.* Find the first $\delta_0 > r/\eta^2$ such that, setting $u(x, 0)$ and $u_t(x, 0)$ as in (3.11) with $\alpha_j = \delta_0$, $\alpha_n = r$ for $n \neq j$, condition (3.12) holds. Set $\delta = \delta_0$.
- *Step 3.* Define $EN = E_j(a)$ as $\mathcal{E}(U_j^A)$, as discussed after (3.19), with $\alpha_j = \delta$; if $EN > EN_{\max}$, go to Step 8, otherwise set $BD = \delta/10$, compute $T_W(\delta)$ through formula (3.40) and go to Step 4.
- *Step 4.* Fix $T > 2 \lim_{\delta \to 0^+} T_W(\delta)$ (so that $T > 2T_W(\delta)$ for every $\delta > 0$).
- *Step 5.* Numerically integrate (e.g., via the "NDSolve" procedure) the finite-dimensional system (3.18) on the interval $[0, T]$. Check whether there exist $k \in \{0, 1, \ldots, N - 1\}\setminus\{j\}$ and $\tau \in (2T_W(\delta), T)$ for which $|\phi_k(\tau)| = BD$: if yes, proceed to Step 6, if no, proceed to Step 7.
- *Step 6.* For every k as in Step 5, evaluate

$$M = \max_{t \in [0, \tau/2]} |\phi_k(t)|$$

and check if the ratio $BD/M > 1/\eta - tol$: if yes, compute $E_j(a)$ via the expression in (3.19) and proceed to Steps 8–9. Otherwise, set $BD = M/\eta$ and repeat Step 5.
- *Step 7.* Increase δ by $step$ and repeat Steps 3–5.
- *Step 8.* If $j \leqslant N - 2$, increase j by one unit and perform Steps 2–5, otherwise go to Step 9.

– *Step 9.* Find $\mathbb{E}_{12}(a)$ as in (3.20); then change the value of a and repeat Steps 2–8.

For our experiments, we chose a ranging from 0.1 to 0.9 with a step of 0.1, together with the remarkable values $a = 1/3$, $a = 14/25 = 0.56$, $a = 2/3$, and

$$N = 12, \ j \in \{0, \ldots, 11\}, \ \eta = 0.1, \ tol = 10^{-8}, \ step = 0.01, \ r = 0.01.$$

We integrated the corresponding system on the interval $[0, 16]$, after having checked that the condition $16 > 2T_W(\delta)$ was always fulfilled. Notice that the presence of $EN_{\max}$ in the above algorithm is only needed to ensure that the algorithm ends; however, in our experiments we continued increasing δ until we found instability and this process always had success. Possible loop cycles in controlling instability, due to an insufficient machine precision, have been bypassed by further slightly increasing the initial value δ by $step$. As for Definition 4.2, if the torsional component is initially set equal to r, of course in Step 2 it has to be taken $BD = r/\eta$.

References

1. Ammann OH, von Kármán T, Woodruff GB (1941) The failure of the Tacoma Narrows Bridge. Federal Works Agency
2. Bartoli G, Spinelli P (1993) The stochastic differential calculus for the determination of structural response under wind. J Wind Eng Ind Aerodyn 48:175–188
3. Benci V, Fortunato D, Gazzola F (2017) Existence of torsional solitons in a beam model of suspension bridge. Arch Ration Mech Anal 226:559–585
4. Berchio E, Gazzola F (2015) A qualitative explanation of the origin of torsional instability in suspension bridges. Nonlinear Anal TMA 121:54–72
5. Brownjohn JMW (1994) Observations on non-linear dynamic characteristics of suspension bridges. Earthq Eng Struct Dyn 23:1351–1367
6. Cazenave T, Weissler FB (1996) Unstable simple modes of the nonlinear string. Quart Appl Math 54:287–305
7. Dickey RW (1970) Free vibrations and dynamic buckling of the extensible beam. J Math Anal Appl 29:443–454
8. Ferreira V, Gazzola F, Moreira dos Santos E (2016) Instability of modes in a partially hinged rectangular plate. J Differ Equ 261:6302–6340
9. Garrione M, Gazzola F (2020) Linear theory for beams with intermediate piers. Commun Contemp Math
10. Gasparetto C, Gazzola F (2018) Resonance tongues for the Hill equation with Duffing coefficients and instabilities in a nonlinear beam equation. Commun Contemp Math 20:1750022 (22 pp)
11. Holubová G, Matas A (2003) Initial-boundary value problem for nonlinear string-beam system. J Math Anal Appl 288:784–802
12. Holubová G, Nečesal P (2010) The Fučik spectra for multi-point boundary-value problems. Electron J Differ Equ Conf 18:33–44
13. Irvine HM (1981) Cable structures. MIT Press Series in Structural Mechanics, Massachusetts
14. Lazer AC, McKenna PJ (1987) Large scale oscillating behaviour in loaded asymmetric systems. Ann Inst H Poincaré Anal Non Linéaire 4:244–274
15. Lazer AC, McKenna PJ (1990) Large amplitude periodic oscillations in suspension bridge: some new connections with nonlinear analysis. SIAM Rev 32:537–578

16. Luco JL, Turmo J (2010) Effect of hanger flexibility on dynamic response of suspension bridges. J Eng Mech 136:1444–1459
17. Marchionna C, Panizzi S (2016) An instability result in the theory of suspension bridges. Nonlinear Anal TMA 140:12–28
18. McKenna PJ, Tuama CÓ (2001) Large torsional oscillations in suspension bridges visited again: vertical forcing creates torsional response. Am Math Mon 108:738–745
19. Smith FC, Vincent GS (1950) Aerodynamic stability of suspension bridges: with special reference to the Tacoma Narrows Bridge, Part II: mathematical analysis, investigation conducted by the Structural Research Laboratory, University of Washington–Seattle: University of Washington Press
20. Yakubovich VA, Starzhinskii VM (1975) Linear differential equations with periodic coefficients. Wiley, New York (Russian original in Izdat. Nauka, Moscow, 1972)
21. Zhukovskii NE (1892) Finiteness conditions for integrals of the equation $d^2 y/dx^2 + py = 0$ (Russian). Math Sb 16:582–591

Chapter 5
Final Comments and Perspectives

Abstract The last chapter of the book contains some final comments and open questions, motivating possible extensions of the presented results in different research directions.

Keywords Multiple piers · Asymmetric side spans · Full plate models · Stretching energy · Stability analysis

5.1 Conclusions

In the first part of the present book, Chap. 2, we have recalled all the basic tools for the analysis of hinged beams with two piers, taken from [6]. The functional and variational setting, as well as the spectral analysis, highlighted that *each pier reduces by one the dimension of the functional space and inserts a Dirac delta within the equation.* Of particular relevance for the rest of the book and for future developments is the behavior of the eigenvalues of (2.8) and the nodal properties of the related eigenfunctions as the position of the piers varies, features that are essential also for engineers in order to study the oscillations of a bridge, see once again Fig. 2.2. Moreover, we showed that the functional space (of codimension two) does not allow to view the fourth order eigenvalue problem (2.8) as the "squared" second order eigenvalue problem (4.2), a fact that highlights how *the stretching energy propagates in a disordered way across the piers.*

In Chap. 3 we reached two main conclusions: the best stability performances are obtained when the piers are placed in the physical range (1.1) and the nonlinear term in (3.1) that better describes the behavior of actual structures is $\|u\|_{L^2}^2 u$. These conclusions were reached after a lengthy and delicate analysis of each nonlinearity and after studying the stability of the prevailing modes (Definition 3.2) from several points of view. At each stage of our analysis, we kept an eye on what was observed by the witnesses of the TNB collapse [1]. This led us to use a suitable notion of stability (Definition 3.4) and to focus our attention on the particular class of solutions having a prevailing mode. Also the conclusion on the "best" nonlinearity takes into account the witnesses of the TNB collapse, in particular the fact that *the system should have small physiological transfers of energy between modes*. The nonlinearity $\|u\|_{L^2}^2 u$ satisfies this feature and has a nonlocal behavior as in real structures.

These results for nonlinear beams turned out to be of crucial importance for the analysis of a more realistic and involved model, the so-called fish-bone model, studied in Chap. 4, which better describes the behavior of a suspension bridge. Indeed, for this model the theoretical tools have limited strength and, if one wishes to have an idea of the phenomena governing its stability, then one should have in mind what happens for simpler models where the results are much more precise. We were able to show the impact of the two main nonlinear forces acting on the deck of a bridge, the restoring forces due to the sustaining cables and to the hangers. We saw that odd and even longitudinal modes respond differently to the slackening of the hangers and that an increment of the elastic constant ς plays against linear torsional stability.

Overall, our results enable us to give some suggestions for the designers of future bridges:

- design bridges with piers in the physical range (1.1);
- avoid the use of "slipping" decks or cables across the piers since we saw that the stretching energy mixes all the modes and makes the stability analysis much more difficult;
- avoid the use of too elastic hangers.

In the next subsections we list some problems left open in the present book and some possible future developments towards a better understanding of the instability in suspension bridges.

5.2 Some Open Questions

- Definition 3.2 leaves some arbitrariness in the choice of α_n ($n \neq j$) in (3.12). In all our experiments, we took $\alpha_n = 0.01$ for $n \neq j$, regardless of the value of α_j. It would be interesting to analyze how the energy threshold depends on the choice of the α_n's; our conjecture is that if we increase the initial value of the residual components, then the energy threshold of nonlinear instability will also increase (recall that condition (ii) in (3.15) needs to be satisfied) but the hierarchy between modes will be maintained. Clearly, this *does not* mean that one should have larger residual modes to improve stability! The important feature is instead that the "weakest" couple of modes would not depend on how the α_n's are chosen, this being the most important response for the stability analysis; in fact, the detection of this couple may help a designer improve the performances of the structure.
- It would be interesting to study (3.25) with $\gamma_1\gamma_3 > 0$. How does the instability diagram look like? Maybe something in between Figs. 3.2 and 3.3?
- How to analytically prove (3.57)?
- The stability analysis performed in Sect. 4.3.2 should be completed with the remaining cases. What happens in presence of multiple torsional eigenfunctions and what happens for $a < 1/2$? Torsional modes other than the second should be

considered, as well. Let us also point out that a full analysis of the Hill equation (4.16) appears mathematically quite challenging. Assume that W_λ solves (3.36), see (3.38), and consider the equation

$$\ddot{\xi}(t) + \left(\kappa^2 + \gamma\,W_\lambda(t)^2\right)\xi(t) = 0$$

for some $\gamma > 0$. The case $\gamma = 1$ is fully described by Proposition 3.4, while the case $\gamma \in (1, 3)$ is partially described by Theorem 4.2. What happens for other values of γ? In the proof of Theorem 4.2 we have seen that the behavior at infinity of the resonant tongues is governed by the Cazenave-Weissler intervals (3.46), but what about the behavior for δ small?

- The stability analysis performed in Sect. 4.3.2 leaves many unanswered questions. A complete description of the behavior of the optimal energy threshold on varying of ς is a challenging task. From what we saw in our numerical experiments, the general trend seems to be that, for ς increasing, the amplitude threshold of linear stability decreases, while much more investigation is needed to clarify the behavior of nonlinear instability. A full quantitative description of these phenomena is still missing.

- Theorem 4.3 leaves several questions open. First, we remark that if the eigenfunctions η_κ are different from those in Theorem 4.1 it may be that $B_\varsigma(w) \neq 0$ also if e_λ is odd: indeed, consider the second eigenfunction e_1 of (2.8) (which is odd) and an eigenfunction of the kind $\mathcal{D}_\kappa + \mathcal{P}_\kappa$ of (4.2) (which is neither odd nor even). Second, Theorem 4.3 *does not* hold if $\eta_\kappa = \mathcal{D}_\kappa$ or $\eta_\kappa = \mathcal{P}_\kappa$, see Theorem 4.1. To see this, take $a = 1/5$ and $e_\lambda(x) = \eta_\kappa(x) = \cos(5x/2)$ so that $\lambda = \kappa = 5/2$. With this choice, we have $B_\varsigma(w) \equiv 0$. We have here exploited the fact that e_λ has "an integer number of periods" (one in this case) on the side span. In fact, if $a \notin \mathbb{Q}$ this cannot happen and we believe that Theorem 4.3 holds also for η_κ not vanishing on the side spans. We leave these as open problems.

5.3 Future Developments

- In this book we have only focused on the structural aspect of the instability problem. The next step is to take into account the fluid-structure interaction [5, 11] and to insert into the model suitable damping effects. We do not expect the aerodynamics to modify the response on the optimal position of the piers but it will probably modify the *quantitative response* in terms of energy thresholds.

- Motivated by the design of most bridges, we have considered here the case of symmetric side spans. The main advantage of this restriction is that one can deal with even and odd eigenfunctions, see the discussion in Chap. 2. But some suspension bridges, such as the three Kurushima Bridges (see [10, Fig. 2.4.6]) have asymmetric side spans. In this respect, let us also recall a forgotten suggestion from the 19th century: while commenting the collapse of the Brighton Chain Pier (1836),

Russell [15] claims that *the remedies I have proposed, are those by which such destructive vibrations would have been rendered impossible*. His remedies were to alter the place of the cross bars and to put stays below the bridge which should be put *at distances not perfectly equal*. His scope was to break symmetry in the longitudinal oscillating modes of the deck. Therefore, the optimal position of the piers should also be discussed in the asymmetric framework, it is not even clear if symmetry yields better stability performances: a full analysis and the comparison with the case of symmetric side spans appear quite important and challenging.

- Several suspension bridges, such as the San Francisco-Oakland Bay Bridge (see [13, Fig. 15.10]), have more than two intermediate piers. Some of the results in the present book may be extended to the case of multiple intermediate piers. With the same proof as for [6, Theorem 1], one can show that the subspace of $H^2 \cap H_0^1(I)$ with n interior vanishing constraints has codimension n. Also Theorem 2.2 continues to hold with some obvious changes. However, a different and general procedure seems necessary to prove smoothness of a solution, since problem (2.6) has exactly the same number of constraints (four) as the order of the ODE. But the main difficulty is certainly to determine the optimal length of the secondary spans in order to minimize the dangerous energy exchanges within the main span.

- The opportunity to introduce full plate models should be evaluated. From Ventsel-Krauthammer [16, Sect. 1.1] we learn that plates may be classified according to the ratio between width and thickness. By taking these values from the Report [1], we deduce that *the collapsed TNB may be considered as a thin plate*, see [7, Sect. 5.2.1] for a detailed discussion. The behavior of rectangular thin plates subject to a variety of boundary conditions is studied in [4, 8, 9, 12]. The role and the optimal position of the intermediate piers could be studied also for thin plate models since they appear more appropriate for decks with a large width.

- The nonlocal behavior of structures such as suspension bridges may intervene also in the differential operators and in other terms, for instance dampings. The stability analysis, both for beams and degenerate plates, appears quite challenging in general frameworks. A possibility would be to compare the results in the present book with results obtained for different beam and plate models with Woinowsky-Krieger-type nonlinearity, such as the ones considered in [2, 14]. For these equations, a full analysis of the nonlinear stability appears extremely difficult and this is a further reason why one should always keep an eye on linear stability.

- In plates without piers, the stretching effects can be controlled, see [3]. As already mentioned, in presence of the piers the stability analysis is much more complicated, see Sect. 3.5. Therefore, a possible suggestion is to design the entire structure in such a way that the stretching effects in a point remain confined to the span to which it belongs. This means that, instead of (3.64), one should study the stability for the nonlinear nonlocal equation

$$u_{tt} + u_{xxxx} - \left(\|u_x\|_{L^2(I_-)}^2 \chi_- + \|u_x\|_{L^2(I_0)}^2 \chi_0 + \|u_x\|_{L^2(I_+)}^2 \chi_+ \right) u_{xx} = 0 \quad x \in I, \ t > 0,$$

that is, an equation where the stretching effects act separately on each span. Here, χ_-, χ_0, χ_+ denote the characteristic functions of I_-, I_0, I_+.

References

1. Ammann OH, von Kármán T, Woodruff GB (1941) The failure of the Tacoma Narrows Bridge. Federal Works Agency
2. Autuori G, Pucci P, Salvatori MC (2009) Asymptotic stability for nonlinear Kirchhoff systems. Nonlinear Anal Real World Appl 10:889–909
3. Bonheure D, Gazzola F, Moreira dos Santos E (2019) Periodic solutions and torsional instability in a nonlinear nonlocal plate equation. SIAM J Math Anal 51:3052–3091
4. Braess D, Sauter S, Schwab C (2011) On the justification of plate models. J Elast 103:53–71
5. Chueshov I, Dowell EH, Lasiecka I, Webster JT (2016) Nonlinear elastic plate in a flow of gas: recent results and conjectures. Appl Math Optim 73:475–500
6. Garrione M, Gazzola F (2020) Linear theory for beams with intermediate piers. Commun Contemp Math
7. Gazzola F (2015) Mathematical models for suspension bridges. Vol 15, MS&A, Springer
8. Grunau H-Ch (2009) Nonlinear questions in clamped plate models. Milan J Math 77:171–204
9. Grunau H-Ch, Sweers G (2014) A clamped plate with a uniform weight may change sign. Discrete Contin Dyn Syst Ser S 7:761–766
10. Jurado JA, Hernández S, Nieto F, Mosquera A (2011) Bridge aeroelasticity, sensitivity analysis and optimal design. WIT Press, Southampton
11. Lasiecka I, Webster JT (2016) Feedback stabilization of a fluttering panel in an inviscid subsonic potential flow. SIAM J Math Anal 48:1848–1891
12. Nazarov SA, Stylianou A, Sweers G (2012) Hinged and supported plates with corners. Zeit Angew Math Physik 63:929–960
13. Podolny W (2011) Cable-suspended bridges. In: Brockenbrough RL, Merritt FS (eds) Structural steel designer's handbook: AISC, AASHTO, AISI, ASTM, AREMA, and ASCE-07 design standards, 5th edn. McGraw-Hill, New York
14. Pucci P, Saldi S (2017) Asymptotic stability for nonlinear damped Kirchhoff systems involving the fractional p-Laplacian operator. J Differ Equ 263:2375–2418
15. Russell JS (1841) On the vibration of suspension bridges and other structures; and the means of preventing injury from this cause. Transactions of the Royal Scottish Society of Arts 1
16. Ventsel E, Krauthammer T (2001) Thin plates and shells: theory: analysis, and applications. CRC Press, Boca Raton

Printed in the United States
By Bookmasters